Praise for

THE ALLURE OF THE MULTIVERSE

"Befitting a discussion about the multiverse, Paul Halpern's latest book contains multitudes. Covering topics from quantum mechanics to general relativity, from the subatomic realm to the edge of the universe (and beyond!), this fascinating and engaging book will expand your mind into new dimensions."

—James Kakalios, author of
The Amazing Story of Quantum Mechanics

"A rich and rewarding history of one of the most astounding ideas in physics and astronomy: the multiverse. Don't believe that other universes might be residing right beside our own, hidden from view in other dimensions? Halpern will convince you to take a second look."
—Marcia Bartusiak, author of *Black Hole*

"The implications of quantum mechanics (a scientific theory that has passed every experimental test of it so far devised) are often bizarre—but none more so than its suggestion of a multiverse of parallel worlds. Once found only on the pages of philosophy journals and in the stories of science fiction magazines, the multiverse concept is now taken seriously by no-nonsense physicists. In clear, accessible prose, Halpern's *The Allure of the Multiverse* explains how this evolution in our scientific understanding of reality has occurred."
—Paul J. Nahin, author of *Time Machine Tales*

"The multiverse is a staple in today's science fiction, but as Halpern shows us, the idea of a multitude of universes has a rich history, with scientists and philosophers debating its merits for centuries. From Friedrich Nietzsche's musings on 'eternal return' to Hugh Everett's mind-bending interpretation of quantum mechanics, *The Allure of the Multiverse* blends history and physics while it both provokes and entertains." —Dan Falk, author of *In Search of Time*

"In *The Allure of the Multiverse*, Halpern takes a deep dive into the history, philosophy, and personalities behind this strangest of ideas. An informative and entertaining read, whichever corner of the multiverse you inhabit." —Will Kinney, author of *An Infinity of Worlds*

"As soon as you contemplate, 'What if?' you enter the multiverse. Halpern's *The Allure of the Multiverse* serves as a masterful guide. Whether you're a curious novice or a seasoned science enthusiast, this book offers a cogent exploration of the maddening infinities our minds can create."
—Chris Ferrie, author of *Where Did the Universe Come From? And Other Cosmic Questions*

THE ALLURE OF THE MULTIVERSE

ALSO BY PAUL HALPERN

Flashes of Creation: George Gamow, Fred Hoyle, and the Great Big Bang Debate

Synchronicity: The Epic Quest to Understand the Quantum Nature of Cause and Effect

Einstein's Dice and Schrödinger's Cat: How Two Great Minds Battled Quantum Randomness to Create a Unified Theory of Physics

The Quantum Labyrinth: How Richard Feynman and John Wheeler Revolutionized Time and Reality

THE ALLURE OF THE MULTIVERSE

EXTRA DIMENSIONS, OTHER WORLDS,
AND PARALLEL UNIVERSES

PAUL HALPERN

BASIC BOOKS
New York

Copyright © 2024 by Paul Halpern

Cover design by Emmily O'Connor
Cover images © Matis75 / Shutterstock.com; © Wirestock Creators / Shutterstock.com
Cover copyright © 2024 by Hachette Book Group, Inc.

Hachette Book Group supports the right to free expression and the value of copyright. The purpose of copyright is to encourage writers and artists to produce the creative works that enrich our culture.

The scanning, uploading, and distribution of this book without permission is a theft of the author's intellectual property. If you would like permission to use material from the book (other than for review purposes), please contact permissions@hbgusa.com. Thank you for your support of the author's rights.

Basic Books
Hachette Book Group
1290 Avenue of the Americas, New York, NY 10104
www.basicbooks.com

Printed in the United States of America

First Edition: January 2024

Published by Basic Books, an imprint of Perseus Books, LLC, a subsidiary of Hachette Book Group, Inc. The Basic Books name and logo is a trademark of the Hachette Book Group.

The Hachette Speakers Bureau provides a wide range of authors for speaking events. To find out more, go to hachettespeakersbureau.com or email HachetteSpeakers@hbgusa.com.

Basic books may be purchased in bulk for business, educational, or promotional use. For more information, please contact your local bookseller or the Hachette Book Group Special Markets Department at special.markets@hbgusa.com.

The publisher is not responsible for websites (or their content) that are not owned by the publisher.

Print book interior design by Bart Dawson.

Library of Congress Cataloging-in-Publication Data

Names: Halpern, Paul, 1961– author.
Title: The allure of the multiverse : extra dimensions, other worlds, and parallel universes / Paul Halpern.
Description: New York : Basic Books, 2024. | Includes bibliographical references and index.
Identifiers: LCCN 2023019652 | ISBN 9781541602175 (hardcover) | ISBN 9781541602182 (ebook)
Subjects: LCSH: Multiverse. | Cosmology.
Classification: LCC QB981 .H247 2024 | DDC 523.1—dc23/eng/20231011
LC record available at https://lccn.loc.gov/2023019652

ISBNs: 9781541602175 (hardcover), 9781541602182 (ebook)

LSC-H

Printing 3, 2024

*Dedicated to the memory of David Zitarelli,
an outstanding teacher, mentor,
and historian of mathematics*

If the doors of perception were cleansed then everything would appear to man as it is: Infinite.
For man has closed himself up, till he sees all things thro' narrow chinks of his cavern.
—William Blake (*The Marriage of Heaven and Hell*)

CONTENTS

Introduction: When One Universe Is Not Enough	1
Chapter One: Eternity Through the Stars	27
Chapter Two: Theories from Another Dimension	59
Chapter Three: Showdown in Hilbert's Hotel	81
Chapter Four: Order from Chaos	127
Chapter Five: Burgeoning Truths	167
Chapter Six: Tangled Up in Strings	191
Chapter Seven: Seasons of Rebirth	219
Chapter Eight: The Time Travelers Party	241
Conclusion: The Reflecting Pool and the Sea	259
Acknowledgments	277
Further Reading	279
Notes	283
Index	291

I think we have enough tsuris with one Verse.[1]
—Stanley Deser, award-winning theoretical physicist

INTRODUCTION

When One Universe Is Not Enough

In a society with omnipresent cameras, seeing is an essential part of believing. Messages are stamped and certified with the watermarks of pictorial proof. "Pics or it didn't happen," states a popular meme. In these days of photographic manipulation not every image is genuine, but authentic photos continue to convey a certain legitimacy.

No wonder the launch of the James Webb Space Telescope (JWST) on Christmas Day 2021 and its ever-growing cache of stunning images released since July 2022 have proven so exciting. Faint, distant galaxies from the nascent era of the universe, only several hundred million years after the Big Bang, have suddenly come to life. Nurseries for stellar formation sparkle and glisten like dew-speckled, blazing wildflowers. No simulations or theoretical equations could match in the public eye such vivid photo evidence—albeit converted from infrared radiation to colorful visible portraits. Space pics, and it did happen!

Given the desire for visualizable signals—by telescope or other means—to cement reality, physicists' growing interest in a multiverse—including unobservable enclaves—seems, at face value, to make little sense. The time-tested scientific method requires experimental verification. Yet the very idea of a multiverse—supplementing the discernable universe with realms beyond direct detection—seems antithetical to the goal of testability. Would any detective draw a conclusion about a possible crime if there were absolutely no possibilities for gathering evidence, such as access to the scene, eyewitness testimony, and so forth?

Moving beyond the potentially observable, therefore, which multiverse theories suggest, seems a radical step, not to be taken lightly. Why not just stick to the measurable, and map out the recordable using powerful instruments such as the JWST? Undoubtedly, within the bounds of telescopic observation, there remains so much out there left to explore.

By instinct and tradition, humankind has sought to understand its environment as thoroughly as possible—to ward off dangers, embrace beneficial opportunities, and make helpful predictions. Charting the world—and, beyond that, the cosmos—out to the very frontiers of detection, following the pathways of the great voyagers, has been a key part of our heritage. By carefully documenting what we find, we strive to fill in the gaps in our comprehension. Maps, once complete, offer the comfort and benefits of knowing all that is out there.

Yet, paradoxically, in scoping out our territory, we also become aware—like fenced-in animals—of the impossibilities. Our curiosity knows no bounds. Any map or system claiming to describe everything begs the questions, "Might there be something else?" and, if so, "Could we somehow access those regions beyond?"

Multiverse models appeal to that sense of wonder. Our imaginations spawn countless alternatives, many beyond the threshold of testability. Fascination with alternate histories and curiosity about worlds unknown has driven public enthusiasm for recent films and television series with multiverse motifs, such as the Academy Award–winning movie *Everything Everywhere All at Once*, the highly watched streaming series *The Man in the High Castle*, and numerous Marvel Cinematic Universe projects. In the popular television series *Rick and Morty*, the titular characters journey, in nearly every episode, to various offbeat parallel universes, sometimes encountering bizarre alternative versions of themselves, such as megalomaniac Ricks and Mortys. Such otherworldly adventures stem from a long-standing literary tradition. Breaking physical barriers, such as limits in space and time, has been a mainstay of science fiction for years.

INTRODUCTION: WHEN ONE UNIVERSE IS NOT ENOUGH

For serious scientists to consider multiverse notions, however, requires far more than fanciful ruminations about uncharted regions and unrealized possibilities. There needs to be a strong explanatory benefit that overcomes the stark disadvantage of a lack of direct detectability. In general, multiverse models offer virtually unlimited mathematical and/or conceptual frameworks upon which the observable features of the universe might be justified, like the enormous, unseen concrete foundations underlying many skyscrapers to support their sleek, lofty structures.

Take, for example, the physics community's ardent pursuit of a simple, unified explanation of the natural forces. Its goal is to express gravitation, the strong nuclear force, and the electromagnetic and weak interactions using the same basic language. One hitch is that gravitation, unlike the other forces, defies conventional attempts to integrate it with quantum physics. In order to accommodate its resistance to standard methods, superstring theory, based on energetic, vibrating strands of energy, has emerged as the leading unification proposal. For reasons of mathematical consistency, the theory makes sense only if housed in a high number of dimensions, typically ten or eleven. Typically, through a mathematical process called compactification, the extra dimensions—beyond ordinary space and time—become curled up into immeasurably minuscule balls or knots. In some variations, they are large, but inaccessible to matter and light, and thereby unseen. Thus, in essence, superstring theory and related higher-dimensional unification attempts apply realms beyond direct detection in order to craft mathematically rigorous, unified descriptions of the natural forces. Assuming such a model someday grasps the trophy of unity, many physicists might find elegant explanation enough, and discount the need for testability and falsifiability of its hidden elements.

If you live in a plush suite in a high-rise building in Chicago, and marvel that it is exceptionally stable in high winds, you are not going to complain that you can't explore the bedrock under its basement. Similarly, many theorists are willing to accept unobservable components in a multiverse model if it supports a promising way of explaining the basic facts of the reality we experience. Just as there are clashing preferences in architecture, though, there is a wide range of opinions and tastes about how seriously to take multiverse schemes.

On one end of the spectrum lies absolute realism—demanding photographic proof or its equivalent to support any claim. The universe, in that view, should be as ironclad as a perfect engine, with all parts labeled and functioning in mechanical precision. Such is the heritage of Isaac Newton, and the clockwork cosmos he delineated. From that point of view, a multiverse is blind faith rather than trusted science.

On the other end is the concept of a "landscape" that embraces every conceivable option. As weird as it sounds, entire universes would exist that we'd never have access to, yet would be as legitimate as ours. The presence of the other universes would be used to help bolster a comprehensive theory of ours. In that case, why would we be in this one, and not in one of countless others? Might there be a selection mechanism that singles out our own enclave as being the best suited for the emergence of intelligent life—an "Anthropic Principle" (as it is called) explaining why we are here by ruling out lifeless alternatives? Or could our presence in this particular universe simply derive from the vagaries of chance—our cosmic abode being a windswept tumbleweed in a desert of meaninglessness?

Extreme caution, outrageous fancifulness, or somewhere in between—such is the range of opinions in the physics community today. A brilliant idea for some might be sheer folly to others, according to taste and tolerance. Without consensus, every funding request for a research project dedicated to indirect tests of multiverse ideas potentially raises battle cries. Yet, a theory explaining nature's workings in a unified way that includes only directly testable assumptions seems further away than ever before. If we don't abandon the key theoretical goal of unification, we may need to compromise—and reconcile conflicting opinions on where to draw the line.

INTRODUCTION: WHEN ONE UNIVERSE IS NOT ENOUGH

The boundaries in contemporary physics between mainstream and far-flung ideas have shifted significantly throughout the years. What is fringe sometimes slips into vogue, and the converse. For instance, before relativity, only a handful of scientists took seriously the notion of a fourth dimension. Now it is standardly applied as a way of including time along with space in amalgamated space-time.

Considering such turnabouts, it seems best, therefore, to remain cautiously open-minded about various multiverse schemes—rather than dismissing them outright. One of my goals in writing this book is to demonstrate how the fluidity of certain physical notions—in some cases transforming a concept that appears unbearably strange into something that seems eminently sensible—suggests avoiding rendering blanket judgments about multiverse ideas. Between unbridled enthusiasm and brutal dismissal lies ample space for thoughtful appraisals of the costs and benefits of various proposals.

QUANTUM WEIRDNESS AND ZOMBIE CATS

Given the scientific-method tradition of confirming every theory through experimental testing, complete realism might seem the most practical approach. However, nature has not made that easy. While in the eighteenth and nineteenth centuries, Newtonian physics, also known as classical mechanics, offered the promise of being able to track the trajectories of any object one chooses—and thus theoretically map out the paths of all things in the observable cosmos—developments in the early twentieth century led the mainstream physics community to abandon the notion that everything is measurable at all times.

In quantum mechanics, which emerged in the mid-1920s, German physicist Werner Heisenberg's uncertainty principle negates that very possibility. Strangely enough, it maintains that certain pairs of physical properties, such as the position and momentum (mass times velocity) of an elementary particle, are such that the more precisely one of them is known the fuzzier the other one becomes. If an experimenter desires exact results, he or she thereby needs to make a choice about which property to measure.

Photographers often need to decide if either the foreground or the background of a scene comes into absolute focus. In some cases, it is impossible to record both exactly in the same high-resolution image. If there is a single photo of an event, and the most important part of it is blurry, pictorial evidence goes out the window. Luckily, often multiple, near-simultaneous photos are taken, providing a complete record for the "pics or it didn't happen" crowd.

In quantum physics, there is no such luxury. Even with the best instruments, experimentalists cannot measure the exact location and precise momentum of a particle at the same time. Moreover, in complex interactions, as renowned American physicist Richard Feynman showed, particles might take multiple routes simultaneously in traveling from one point to another, an idea called "sum over histories." Unlike classical physics, in which each object travels along a single, predictable trajectory, in Feynman's scheme, a particle's overall behavior emerges from an array of different paths, each weighed with different probabilities. We witness the overall outcome, not the alternate histories that went into it. Fundamentally, therefore, the world we see contains only a fraction of all information about its potential characteristics. The complete set of data, called quantum states, resides in an abstract realm of unlimited dimensions—which Hungarian-American mathematician John von Neumann dubbed "Hilbert space."

In line with the philosophical musings of modern physics titan Niels Bohr, in the late 1920s von Neumann described a two-step procedure for quantum processes. Widely adopted, it has become known as the "Copenhagen interpretation," in honor of the Danish city where Bohr, at his institute, gathered some of the most prominent thinkers dedicated to quantum physics. It is also called the "orthodox interpretation."

In the first phase of von Neumann's process, quantum states evolve according to objective, deterministic relationships, albeit within

INTRODUCTION: WHEN ONE UNIVERSE IS NOT ENOUGH

Hilbert space rather than the tangible world. Describing such developments is relatively straightforward.

In phase two, however, he introduced a very peculiar role for human observers. By taking measurements of a particular kind—for instance, position—observers would cause a complex quantum state embracing a range of possible positions to "collapse," with certain likelihoods of outcome, into one of its simpler components. The state would topple like an intricate house of cards down to a narrow pile, resulting in a single value of the measured property—for instance, the pinpoint location of an electron. Weirdly enough, if a different type of measurement had been chosen—momentum instead of position, let's say—the comprehensive quantum state would offer a blend of momentum possibilities and collapse, upon measurement, into one of those options. Thus, quantum mechanics, according to the Copenhagen interpretation, depends on conscious observation to single out a particular property and narrow down its value.

As Erwin Schrödinger, Albert Einstein, and many others have pointed out, one major issue with that interpretation is the artificial division between the observed and the observer. After all, human observers are fundamentally made of elementary particles too. What gives humankind—or conscious entities in general, able to decide which kinds of physical measurements to take—the special power to initiate a quantum process?

In one of his final lectures, Einstein wondered if even a mouse observing a quantum system could measure a physical property and cause its state to collapse. Why just people? The need for a sentient observer, Einstein felt, was a clear inadequacy of the theory, which needed to be replaced by a more objective mechanism.

Choosing a different animal to make his point, in his famous cat conundrum, Schrödinger brilliantly illustrated some of the tricky dilemmas associated with quantum measurement. Imagine, he wrote, a cat placed in a closed box, along with a radioactive sample with a 50 percent chance of decay during a given time interval, a Geiger counter, a hammer wired to the counter, and a flask of poison. Suppose if the sample does decay, it would set off the counter, trigger the hammer to strike and shatter the flask, release the poison, and kill the cat. On the other hand, if it doesn't decay, the cat would be spared.

According to the standard interpretation of quantum mechanics, the sample would remain in a mixed quantum state of decayed and not-decayed until the box is opened. At that point, a sentient observer would effectively measure that state and cause it to collapse into one of the two possibilities. Similarly, therefore, until the box is open, the poor cat would hypothetically persist in a zombie-like mixed quantum state of dead and alive. Clearly that's absurd, Schrödinger noted, necessitating a more sensible description of quantum processes.

Furthermore, as American physicist John Wheeler and others have pointed out, if quantum mechanics is universal, it should apply to the universe itself. In theory, the cosmos as a whole should be represented by a quantum state of unimaginable complexity. But clearly the universe cannot have an outside observer triggering its overall quantum state to collapse.[2]

Weaving the need for conscious observation into natural processes that have taken place for billions of years seems strange indeed. However, as Bohr once said to Austrian physicist Wolfgang Pauli, in a completely different context:

> We are all agreed that your theory is crazy. The question that divides us is whether it is crazy enough to have a chance of being correct. My own feeling is that it is not crazy enough.[3]

While Bohr, who stubbornly upheld the orthodox quantum view, did not always apply the same standard to himself, another thinker, Hugh Everett, a young PhD student under Wheeler in the 1950s at Princeton, pressed quantum measurement theory to even greater "craziness." In his proposal to remove human intervention from the quantum picture, Everett introduced the first prominent type of multiverse model. The implausibility of the mainstream approach thereby drove the weirdness of the multiverse at its very start.

Everett's imaginative alternative effectively chopped off von Neumann's step two, as if it were the moldy part of an otherwise savory loaf of bread. It posited that quantum states never actually collapse. Rather, there is a "universal wave function" that evolves indefinitely—akin to an ever-flowing river with many persistent branches. Strangely enough, even after measurement, both what is observed and the observers themselves would remain in mixed states, encompassing a range of outcomes and witnesses. Everything would happen smoothly and privately—like isolated movie theaters in a multiplex each projecting a different experience. The replica scientist in each branch would never know about those in the other branches. The universe would simply go on with parallel strands representing each possible outcome woven together into a resilient fabric of reality.

For example, if someone tried to enact—cruelly enough—Schrödinger's cat experiment, there would be no ambivalence. In one branch of reality, the sample would decay, the hapless feline would be poisoned, and one version of the observer would open the box and mourn the loss. In another, the sample would remain intact, the cat would survive, and another, equally real version of the observer would rejoice. Both outcomes would coexist in the universal quantum state representing reality.

Wheeler sent a version of Everett's thesis to astute gravitational physicist Bryce DeWitt for publication in a journal. Initially, DeWitt raised deep objections to the notion that observers split, arguing that he'd never personally experienced such a sensation. Everett responded that we don't feel Earth rotating either. DeWitt was impressed, became a convert to the idea, and ended up as its principal promoter in the latter decades of the twentieth century. In a paper published in 1971, he dubbed it the "Many-Universes Interpretation of Quantum Mechanics," which would become more commonly known as the "Many Worlds Interpretation (MWI)."[4] Bizarre as the notion of ever-branching universes sounded to a hard-headed physicist such as him, it was far less ludicrous to him than the notion that mere humans—collections of atoms themselves—played a critical role in nature's workings. His thoughtful advocacy of going even "crazier" to explain quantum weirdness in a self-consistent manner sparked broad public fascination with the notion of a multiverse.

A MÉLANGE OF MULTIVERSES

Surprisingly the term "multiverse" did not originate in the world of physics. Rather, it was American philosopher and psychologist William James who introduced the expression in the 1890s as a way of characterizing a morally ambiguous cosmos that doesn't distinguish right from wrong. Around 1970, speculative-fiction writer Michael Moorcock applied the same term in a very different context. He envisioned characters with different avatars in various parallel worlds. Each avatar would share some, but not all, of the personality traits of the overarching character.

When, in the same year, DeWitt first brought the MWI—with its mind-bending prospect of alternative realities populating a universal quantum state—to widespread public attention through an article in *Physics Today*, the physics community had yet to adopt the term "multiverse." The epithet gradually caught on among physicists when, spurred in part by the MWI, interest in parallel universe ideas began to pop up in assorted fields like spring crocuses.

Once physicists began using the expression, its use in popular culture blossomed even further. In the past decade, in particular, the term has surged in popularity.[5] The increasing use of the expression in the Marvel Cinematic Universe, including blockbuster films such as *Spider-Man: No Way Home* and *Doctor Strange in the Multiverse of Madness*, has lifted the idea from a technical construct to a widely shared meme. The critical acclaim and record number of major

INTRODUCTION: WHEN ONE UNIVERSE IS NOT ENOUGH

Oscar wins for *Everything Everywhere All at Once* has undoubtedly bolstered the term's popularity even more. Of course, currently only in fantasy, such as the world of movies, might we imagine characters passing swiftly from one universe to another and encountering (and often fighting) their near-doppelgängers. Science, focused on complex calculations and technical proofs, is far less flashy.

How does one make a multiverse? Let me count the ways. Better yet, let's try to delineate various uses in physics for realms beyond the directly observable—from higher-dimensional spaces to enclaves of the universe with distinct physical properties. Some physicists have strived to classify multiverses into numbered types—notably MIT physicist Max Tegmark's taxonomy involving four levels: two in cosmology, the MWI representing the third, and the set of all mathematical structures rounding out the bunch.[6]

However, any such numerical scheme tends to pave over nuances in the thinking of various physicists about which not-directly-measurable components of theories are acceptable and which are outlandish. Given that modern physics has largely broken with pure, objective realism already, the barriers between normal and intolerable are not always obvious—nor are they universally accepted. What is weird to some might be run-of-the-mill to others—and to yet others, not quite weird enough.

Take, for example, the idea of dimensions. Traditionally speaking, we observe only three—length, width, and height. Most nineteenth-century physicists would have shaken their heads and rolled their eyes upon hearing talk of anything beyond that. At that time, they associated the fourth dimension and beyond with either esoteric mathematics or sham psychic mediums. Indeed, when Einstein proposed the special theory of relativity in 1905—what happens as bodies approach light speed—he framed the effects of such ultrafast travel as contractions along the direction of motion in three-dimensional space, as well as the stretching of time intervals. That is, he kept the two categories, space and time, distinct.

Two years later, however, mathematician Hermann Minkowski cleverly expressed special relativity in a much more natural way by proposing a four-dimensional amalgamation of space and time, called space-time. He reframed the contractions and stretchings as a kind of four-dimensional rotation that takes from space and gives to time—keeping space-time, as a whole, invariant. Within a four-dimensional context, Einstein's Alice in Wonderland–like transformations suddenly made more sense.

Thinking of four dimensions as unnecessarily abstruse, Einstein resisted Minkowski's proposal for several years, until respected colleagues persuaded him that its mathematical formalism actually made his theory simpler, rather than stranger. He would go on to make excellent use of the fourth dimension in the general theory of relativity, published in 1915.

In general relativity, Einstein expressed gravitation as the warping of space-time in the presence of matter and energy. Its curvature takes place along a normally inaccessible extra dimension—similar to how we generally travel along the surface of spherical Earth, and rarely in the direction of its interior.

No matter how much a theory asserts that an additional dimension is impenetrable, however, its mere presence conjures visions of hidden passages and furtive shortcuts. Indeed, Einstein and a collaborator explored such possibilities, which, in the mid-1950s, Wheeler incorporated into the notion of "wormholes": shortcuts through space that connect widely separated regions. In the late 1980s, Wheeler's student, Kip Thorne, further explored the possibility of identifying traversable wormholes that astronauts might safely cross. The wormholes might join two otherwise distant parts of ordinary space, link to different eras of time, or even connect with regions of space-time that would otherwise be wholly separate from ours. Director Christopher Nolan played with that idea in the 2014 film *Interstellar*, for which Thorne served as scientific consultant and co-producer.

INTRODUCTION: WHEN ONE UNIVERSE IS NOT ENOUGH

Conceivably, wormholes might allow for travel to parallel universes, suggesting one variety of multiverse model: the space of all universes connected with ours. Or they hypothetically might permit travel into the past and changes to the timeline of history—spawning yet another multiverse scheme involving alternative realities. For example, a backward time traveler who inadvertently prevented Franklin Roosevelt from ever becoming president might return to an alternative future on a different branch of reality in which the Axis powers defeated the Allies. Backward time travel scenarios remain very theoretical (and perhaps even impossible), for sure, but are discussed, nonetheless, in the pages of serious scientific journals.

Higher dimensions have captivated physicists in another way: offering the prospect of unifying the natural forces, with the aim of bringing gravitation and other interactions within a common mathematical framework. Only a few years after Einstein published the general theory of relativity, Theodor Kaluza, a young mathematics instructor, sent him a paper proposing a way of adding another dimension to allow for the inclusion of electromagnetism—the other then-known fundamental force—along with gravitation. Some time later, Oskar Klein independently developed an equivalent five-dimensional unification model that meshed well with quantum physics. Hence, unification proposals that include extra dimensions are sometimes called "Kaluza-Klein theories." Despite his earlier distrust of higher dimensions, Einstein worked with several different research assistants on his own five-dimensional unified field theory notions, before finally giving up in 1943 and spending the final years of his life pursuing other schemes for unity.

Once again, we might raise Bohr's critique of Pauli. Is unification in five dimensions crazy, or not crazy enough? By the 1970s and 1980s, physicists realized that they'd need to expand their horizons. Two more natural forces—the strong and weak nuclear interactions—had come to their attention. They turned to even-higher-dimensional models to encompass them, along with gravitation and electromagnetism, in unification proposals. Supergravity models of eleven dimensions and superstring models of ten dimensions emerged as contenders for paths to unity. Theorists determined that the high number of dimensions was needed for mathematical rigor—to cancel out certain questionable terms that appear in lower-dimensional models—as well as to encompass all four forces. Hence, in only a few decades, such large dimensionalities evolved, according to the theoretical community's perspective, from being almost laughable to virtually essential.

Superstring theories, and string theories in general, involve replacing point particles on a fundamental level with vibrating strands of energy. The "super" part refers to a hypothetical property of the subatomic world called supersymmetry, in which, at extremely high energies, matter components could transform into force carriers, and the converse. In the 1990s, vibrating membranes entered the picture through an amalgamation of the various models called "M-theory."

Clearly, despite theoretical musings, ordinary space remains three-dimensional, and conventional space-time is four-dimensional. Therefore, in string and M-theories the extra dimensions are typically curled up into tiny balls or knots. Imagine walking along such a curled-up extra dimension, and barely getting anywhere before ending up exactly where you started—in a kind of spatial *Groundhog Day*. Those twisted-up spaces are so minuscule—far, far smaller than the scale we measure with particle colliders—that they cannot possibly be observed. As it turns out, researchers have estimated that there are some 10^{500} (1 followed by 500 zeroes) ways of curling up the extra dimensions, each of which specifies a kind of universe. Rather than home in on a unique vision of how all of the natural interactions emerge from mathematical relationships in a ten- or eleven-dimensional realm, string and M-theories have generated an embarrassment of riches. There is no clear mathematical trick for narrowing the options down to a single theory. Consequently, the numerous possible configurations have led to yet another kind of multiverse, called the "string landscape." It consists of the range of possible universes bearing different physical properties associated with the myriad ways of curling up the extra dimensions—with one of those universes ours, theorists hope.

In little more than a century, the theoretical physics community has evolved from reluctantly accepting time as the fourth dimension to make relativity more straightforward, to embracing a mayhem of string theory scenarios in ten or eleven dimensions that seem to have scant hope of simplification. While some researchers decry the currently confusing state of affairs, others concede that string theory seems the only viable way forward for unification, given the failure of particle-based approaches in the past.

While such a high number of dimensions seems odd, mainstream quantum physicists perform calculations in abstract Hilbert spaces of unlimited dimensions all the time. A key difference, however, is that quantum physicists expect that measurements in Hilbert space (explained via the Copenhagen interpretation or other means) ultimately produce perceptible results in lower-dimensional space. For the string landscape, the process for narrowing down the possibilities to our own tangible reality remains far less certain. String theory thereby currently lacks a recipe for unification—only serving a taste of what could be someday—even though it is based on an age-old cookbook that includes Minkowski's space-time union, geometric relationships in general relativity, five-dimensional Kaluza-Klein theory, and mathematical transformations applied to quantum states in Hilbert space as some of its antecedents.

LANDSCAPES AND DREAMSCAPES

While the notion of a multiverse in physics is relatively recent, the mental construction of alternative worlds is an age-old pursuit. Weaving yarns is a natural process. In dreams, our minds automatically craft strange visions of events that never actually happened or at least transpired differently. Successful planning often involves mentally weighing alternative scenarios, and singling out the optimum. Grandmasters in chess, for instance, might consider numerous chains of potential moves and countermoves—thinking many steps ahead—before even advancing a pawn.

Some philosophers and theologians, in trying to fathom divine thought processes, have imagined a deity similarly pondering every manner of creation before bringing one into existence. For instance, Gottfried Leibniz speculated that God is not only all-seeing and all-knowing about everything in the cosmos but is similarly omniscient about the composition and doings of every conceivable reality. From that set, He has chosen the best of all possible worlds to be ours. The brilliant satirist Voltaire ruthlessly mocked that idea through the chronically sanguine character Pangloss in *Candide*, who distills from any tragedy the rosiest interpretation. Its wit relies on our inclination to see the dark side of history and feel that humanity has been rather unlucky. Yet, compared to every possible cosmic outcome, we must concede that at the very least we are lucky enough to be on a thriving planet with the underlying conditions to support intelligent life.

Multiverses, we see, aren't necessarily extensions of the tangible, physical world. We might divide them broadly into two categories: those that enlarge the universe along physical lines, such as proposing realms beyond observability that are still "out there" somewhere, versus those that exist in a hypothetical domain of possibilities and are used mainly for purposes of comparison. That is, some are landscapes, and others are dreamscapes.

Contemporary physics, in grappling with the questions of "What is real?" and "Why does reality have particular properties?" has considered both varieties—physical extensions and unrealized alternatives. Both arise in Einstein's general theory of relativity, which includes a plethora of finite and infinite solutions for the geometry of the universe. For example, space might be positively curved, like the surface

INTRODUCTION: WHEN ONE UNIVERSE IS NOT ENOUGH

of a sphere; negatively curved, like a saddle shape; or "flat," perfectly straight in all three dimensions, like a box extended forever in every direction. Each of those possibilities—and more—turn up as general relativistic solutions.

In contrast with Newtonian physics, which posits a single, unchangeable grid called "absolute space" through which astral bodies move during a uniform timeline called "absolute time," general relativity embodies astonishing flexibility. Nevertheless, after proposing the theory, Einstein deeply hoped to find physical reasons to guarantee a single, finite, stable cosmic solution.

Much to his dismay, the first solution he developed, based on a positively curved hyperspherical geometry, proved unstable. Attempting to rectify that situation, he added a new stabilizing term to this theory, called the cosmological constant, which counteracted the clumping effects of gravity. That gave him the static result he sought.

However, once mounting evidence for an expanding universe emerged through telescopic investigations of galaxies, Einstein shifted his stance. Along with Dutch scientist Willem de Sitter, in 1932 he proposed a model of the universe that is infinite in extent, expanding indefinitely, and spatially flat. In crafting what became known as the Einstein–de Sitter universe, they set the cosmological constant to zero, removing it from the theory—which no longer required a stabilizing factor. That model formed the conceptual seed of what later became known as the Big Bang theory.

Take a cauldron of scientific curiosity, infuse it with universe models that extend endlessly in all directions, stir in the myriad alternative solutions, and all manner of concoctions arise—landscapes and dreamscapes alike. For example, one such landscape stems from the finiteness of the speed of light, limiting what we could possibly observe. Beyond the zone from which any signals could reach us must almost certainly lie enclaves that forever elude detection. The result suggests an essential multiverse—logically necessary, because it would be nearly impossible to believe that the universe simply stops beyond the horizon of observability.

More abstractly, within the theoretical space of cosmic parameters—such as curvature, smoothness, cosmological constant, and so forth—there would be numerous other options, leading to a more conceptual kind of multiverse. Those might either be dismissed as purely mathematical or be taken seriously as physical alternatives—depending on theorists' preferences. That is, a multiverse composed of alternate solutions to general relativity could be seen as a kind of intellectual dreamscape (conjuring the alternatives and then dismissing them as unphysical) or treated instead as actual contenders within a landscape of genuine options. Theorists' proclivities steer such choices.

In aspiring to create a quantum theory of gravitation, Wheeler favored treating the alternative solutions in general relativity as constituents of an effervescent "geometric foam" that emerges at extremely high energies. Somehow, out of that foam, our simple cosmology emerged as the optimum path through the abstract space of parameters—which, according to Feynman's "sum over histories" approach, represents the "classical" (Newtonian physics) limit. Wheeler's notion sounded fascinating, but never got very far, because of the experimental impossibility of reaching such high energies coupled with the formidable mathematical challenges of constructing viable quantum representations of general relativity (the kinds of difficulties that ultimately drove many theorists to string theory).

Quantum physics aside, even in standard cosmology, questions arise as to how the universe ended up so regular. Flat geometry and isotropic (same in all directions) expansion turn out to be right on the mark—a good call by Einstein and de Sitter. Observations that the growth of the universe is speeding up, however, are consistent with a cosmological constant that is not exactly zero but, rather, a very small positive value. Why so close to zero, theorists wonder, but not quite? Other cosmic oddities include the fact that the entropy, or measure of disorder, of the observable universe must have begun extremely low; otherwise, it would have started with little or no usable energy to create the stars and other thriving astral objects we observe. Finally, many of the constants of nature, such as the strength and range of electromagnetism versus that of the other forces, seem auspicious for the creation of galaxies, stars, and planets.

In 1970, hoping to use the existence of intelligent observers to explain particularly favorable cosmic conditions, Brandon Carter—encouraged by Wheeler—proposed several variations on what he'd dubbed the "Anthropic Principle": the notion that conditions in our region of space-time and/or in the universe itself must be consistent with the eventual emergence of humans (or other intelligent beings) able to observe such favorable properties. The most far-reaching version, the "Strong Anthropic Principle," relies on a multiverse to explain the benign conditions of our universe. Cosmological parameters and conditions would be very different from one universe to another. Ours would be selected as the one universe that could produce the stable stars with planetary systems that support the physical and chemical processes required for intelligent life to thrive. Thus, our mere presence as conscious observers would guarantee that we resided in such a cosmic oasis among the wasteland of alternatives.

Decades later, Carter's hypothesis would be applied to the string landscape in an attempt to narrow down the myriad possibilities. The main selection principle, in that case, would be models that lead to a small, but nonzero, cosmological constant—producing just the right spatial growth to support the eventual emergence of living planets such as ours. A large cosmological constant, in contrast, would have counteracted gravity's ability to clump together material and prevented the formation of galaxies, stars, and planets. Without stable worlds and shining stars, life as we know it never could have developed. The fact of our existence thereby rules out such lifeless universes with sizable cosmological constants, and weeds out the string theory configurations that produce such unfavorable models.

However, back in the 1970s, at the time Carter published his paper, the bulk of physicists were still holding out hope of explaining physical parameters via calculations rather than philosophizing. They anticipated that innovations in the science of the universe would ultimately suffice to resolve all mysteries about its properties.

Indeed, Carter conceded in his article that whenever possible, it would be best to take a purely mechanistic approach and leave humanity's presence out of cosmology. Certain facts, such as the size and density of a hydrogen gas cloud sufficient to coalesce via gravity into a shining stellar body, leading to stars of particular mass ranges, fell into the category of "traditional predictions" based purely on physical constraints. Likely, most theorists reading the article at the time fully agreed with that practical slant.

BUBBLE, BUBBLE, JOY OR TROUBLE?

Indeed, by the close of the 1970s and the start of the 1980s, Alan Guth and others had introduced a variation of the Big Bang theory intended to explain cosmic regularities without necessarily invoking the Anthropic Principle. Dubbed "inflation," Guth's model supposes that the universe went through an extremely brief stage of ultra-rapid expansion very early in its history. Just as pulling a carpet quickly in all directions reduces its wrinkles, the inflationary era, according to its advocates, helped even out any irregularities in the early universe. Such an interval of smoothing helps explain why many of the features of the celestial dome are so consistent in all directions despite their tremendous distances from each other. Smoothing during the spurt of inflation similarly justifies why the universe appears spatially flat, rather than negatively or positively curved.

Strangely enough, shortly after Alan Guth and others introduced the notion of cosmic inflation, Paul Steinhardt, Andrei Linde, and Alexander Vilenkin—who had each developed variations of the theory—separately pointed out that if the observable universe began with inflation, such a process would likely be easy to trigger elsewhere in the cosmos, leading to other inflationary bubbles. In fact, the primordial cosmos would be a bubbling froth of multiple expanding universes—yet another kind of multiverse. In some of these, inflation might just keep going, a situation called "eternal inflation." The alternate universes would be inaccessible today, lying far beyond the reach of observability.

Many advocates of eternal inflation have brought back the Anthropic Principle to explain why we reside in our particular universe. Ironically, a theory originally intended to smooth out the observable universe dynamically (by means of physical processes directly causing changes), without use of a selection principle, now seems to need one to rule out us having ended up in any of the myriad other contending universes with less favorable conditions.

Though such parallel domains would be unreachable at present, in 2010 researchers Hiranya Peiris and Matthew Johnson speculated that the imprint of early collisions between their formative bubbles and those of our own observable universe might have persisted. They proposed analyzing the cosmic background radiation with the aim of identifying such "scars." Their research team identified a few contenders in data gathered by the Wilkinson Microwave Anisotropy Probe (WMAP) satellite, but none with statistical significance. Since then, novel proposals for attempting to find such bubble collisions using polarization (the direction in which photons coil) profiles of the cosmic background radiation have emerged, awaiting possible realization via new surveys. Hence, testing the eternal inflation hypothesis—and thereby one version of the multiverse notion—remains conceivable, but far from a sure thing.

Without even indirect evidence to bolster the various multiverse hypotheses, from the MWI to the string landscape and eternal inflation, they have continued to invite strong criticism from those who insist—rightly or wrongly—that anything lacking the prospect of verification or falsifiability through testing isn't truly science. The urges of multiverse enthusiasts to await comprehensive explanations of the natural world that might lie ahead before dismissing the means to get there does little to allay skeptics' concerns. A deep rift has opened between those willing to include completely unreachable cosmic regions as aspects of theories, and those who find extending physics beyond potentially measurable domains to be utter folly.

To get a taste of the latter, consider the scathing words of writer John Horgan:

> Science is ill-served when prominent thinkers tout ideas that can never be tested and hence are, sorry, unscientific. Moreover, at a time when our world, the real world, faces serious problems, dwelling on multiverses strikes me as escapism—akin to billionaires fantasizing about colonizing Mars. Shouldn't scientists do something more productive with their time?[7]

That debate came to the fore in 2017, when Steinhardt, along with physicists Anna Ijjas and Abraham "Avi" Loeb, rattled the world of science by publishing a sharp critique in *Scientific American* of the notion of an inflationary era. Given that Steinhardt was one of the pioneers of inflation, that critique was especially jolting. The researchers argued that one of the original goals of inflation, to explain why our observable universe looks the way it does, no longer applied. In fact, they argued, inflation theory's implication that other bubble universes likely exist meant that our universe would no longer be unique and that the chances of its special features would reduce to zero. As the team noted: "Because every patch can have any physically conceivable properties, the multiverse does not explain why our universe has the very special conditions that we observe—they are purely accidental features of our particular patch."[8]

Steinhardt further expounded his critique in a 2020 oral history interview:

INTRODUCTION: WHEN ONE UNIVERSE IS NOT ENOUGH

> The issue with the multiverse is, it predicts literally that there are patches of space which manifest every possible conceivable outcome that the laws of physics allow.... The goal of inflation was to explain, among other things, why the universe is spatially flat. But in the multiverse, there's an infinite number of regions which are [negatively curved] and [positively curved].[9]

In launching his critiques of eternal inflation, Steinhardt had another type of cosmological model in mind—with bounces instead of a bang and bubbles. In the early 2000s, he, along with several other physicists including Neil Turok, Justin Khoury, and Burt Ovrut, had developed an alternative to inflation, called, in various incarnations, the "Ekpyrotic Universe" and the "Cyclic Universe," that removed the need for a spatial multiverse by positing recurring cataclysmic events that smoothed out the universe without inflation. However, curiously it introduced the requirement of at least one other parallel universe—separated from us by a fifth dimension—that would periodically crash into ours. Therefore, while it isn't a multiverse in space, one might argue that it is a multiverse in hyperspace.

Moreover, the concept of cycles in time has deep kinship with multiverse notions. Given an unlimited sequence of cosmic eras, it allows for the possibility that events on Earth might someday be repeated. Trillions of years from now, a randomly reassembled version of you might be perusing a replica of this very page.

The notion of endless repetition is hardly new; it pops up repeatedly in the history of ideas. We'll see how nineteenth-century philosopher Friedrich Nietzsche obsessed about the prospect of his entire life, for better or worse, randomly recurring over and over again in what he called "eternal return."

Indeed, Steinhardt's critique that eternal inflation would allow for "every possible outcome" could well be said of a reality with an indefinite (and perhaps even infinite) series of cycles. In many ways, cyclic models—including another alternative called "Conformal Cyclic Cosmology," proposed by Roger Penrose—possess a reliance on unobservable realms similar to multiverse models, albeit in time rather than space.

Linde and others soon penned a rebuttal to the *Scientific American* piece, pointing out the testability of the inflation hypothesis. Indeed, from their perspective, cyclic collisions in an unseen dimension beyond direct observability was the far-fetched model. Better a multiverse in ordinary space, which is subject to known physical law, rather than speculations about higher dimensions, they argued.

The fiery debate rages on. For some, multiverse schemes are perfectly acceptable science. For others, they are all gleam and glitter, without genuine substance. If parts of the cosmos are wholly disconnected, does it make sense even to talk about them? Or might their existence, though only inferred not directly measured, be needed to shed light on our own part of space? As in the famous skirmishes between Einstein and Bohr about quantum mechanics, assertions of realism battle more abstract approaches. In that case, history has declared Bohr the winner. Ultimately, we don't know how it will judge the multiverse clashes of today.

Virginia Trimble, who has written much about the history of astronomy and astrophysics, takes the concept seriously. As she remarked:

> From a historical point of view, every time there has been a one/many controversy (Earth-like things around sun; stars with planet families; galaxies; clusters of galaxies; epochs of star formation—those are not in chronological order), the "many" folks have turned out to be the winners (adjudged more nearly correct over the years). This very much inclines me to "many" here again. The Astronomer Royal [Martin Rees] does take it seriously, and he has always been one of my sources of "ground truth."[10]

INTRODUCTION: WHEN ONE UNIVERSE IS NOT ENOUGH

Indeed, our concept of space and the universe—which, by traditional definition, includes everything that's out there—has changed dramatically over the millennia. Some, but not all, of the ancient Greek philosophers thought that Earth was central, and that the sun, moon, five visible planets (from Mercury to Saturn), and starry dome were kept aloft above us in a realm not particularly far away by modern standards. Eventually, a heliocentric vision of the solar system won out, thanks in part to Galileo's invention of the astronomical telescope in 1609. As astronomers mapped out the vast number of stars in the Milky Way, many began to regard it as the entirety of the universe. It took more than three centuries before a far larger instrument showed that spiral objects once thought to be gas clouds within the Milky Way were actually galaxies in their own right instead, far beyond its periphery.

In tandem with trying to map out the knowable, speculation about realms beyond is one of humanity's most ancient pursuits. In many cases, such as the assertion by sixteenth-century Italian philosopher Giordano Bruno that there are myriad worlds in space—which led, in part, to him being burned at the stake[11]—such ruminations have eventually proven accurate. Vindicating Bruno and others, astronomers have identified many thousands of exoplanets in recent decades and believe that is only the tip of the iceberg. Even as you read these words, the Webb space telescope is being put to good use to find many more—with a good portion of comparable size to Earth. Scientists remain optimistic that habitable planets will eventually be found.

Luckily, belief in a multiverse doesn't draw the wrath of an Inquisition. Nonetheless, until it is testable, at least indirectly, it remains controversial. For many, it has already proven enticing, but will it prove correct? Only time will tell.

At the present time the entire life of our planet, from birth until death, is being detailed day by day with all its crimes and misfortunes on a myriad of brother stars. What we call progress is imprisoned on every Earth, and fades away with it. Always and everywhere in the terrestrial field the same drama, the same décor; on the same limited stage a boisterous humanity, infatuated with its greatness, believing itself to be the universe, and living in its prison as if it were immense spaces, only to soon fall along with the globe that carried—with the greatest disdain—the burden of its pride. The same monotony, the same immobility on foreign stars. The universe repeats itself endlessly and fidgets in place. Eternity infinitely and imperturbably acts out the same performance.

—Louis-Auguste Blanqui, *Eternity Through the Stars*, translated by Mitchell Abidor

CHAPTER ONE

ETERNITY THROUGH THE STARS

Louis-Auguste Blanqui, Friedrich Nietzsche, and the Quest for Replica Worlds

Perched on a small planet, and confined to the present moment as it ambles steadily toward the future, humanity's vantage point within all eternity is humble, isolated, and tenuous, to say the least. Yet we're an audacious lot. Though Earth today is plenty complex—and has always been baffling—our ambitions have always reached much farther than the confines of our rocky stoop.

Before we explore strange visions of the multiverse, our first step is to venture well beyond our home planet and appreciate the sheer vastness and complexity of the observable universe itself. Even in that domain, which used to be considered far smaller and has grown considerably in measured scale over the millennia, thinkers have pondered parallel Earths—worlds similar to our own, assembled through sheer chance.

Space travel is relatively modern and still awfully slow. Human transport has taken us to the moon, but disappointingly not much beyond that. Robot craft—a much safer pursuit—have now reached just beyond the limits of the solar system, but this has scarcely made a dent in our prospects for interstellar travel. While ancient mariners could extend global knowledge by sailing to lands hitherto unknown (by those of their homelands at least), the prospect of spacecraft exploring the immense reaches of the cosmos is still far off.

Fortunately, Earth is bathed in the light of the distant heavens—a tiny fraction of which might be seen by the naked eye, but sufficient nonetheless to galvanize our imaginations and propel them skyward. Telescopes and modern instruments complete the task by concentrating those luminous signals into vivid images, which can be analyzed scientifically. Yet even in ancient times, well before such instrumentation, enough could be seen to inspire thoughts of realms beyond the mundane.

Along with stargazing, finding patterns in nature is an age-old occupation. There are plenty of patterns in astronomy, from the phases of the moon and the rhythms of solstices and equinoxes, to the steady passage of the constellations and the traceable paths of the planets amongst the stellar background, as well as more exotic, but still anticipatable, phenomena such as the advent of comets and the coming of eclipses. Those in ancient days who understood and tracked such astral repetitions lent their skills to rulers as respected authorities—in terrestrial, as well as celestial, affairs—serving as astrologers as well as astronomers. Such sages helped develop elaborate calendars, which in some cultures spanned many thousands, millions, or even billions of years. Consider, for example, the Long Count of the Maya calendar, including cycles lasting almost eight thousand years, and the *kalpa* of the Hindu tradition, which the Puranas scripture describes as a great cycle lasting more than four billion years—in each case involving the creation and destruction of the entire universe.

Greek philosophers debated the essential ingredients of nature, a pursuit that has a deep connection with the concept of cycles. Plato, for instance, spoke of a Great Year of tens of thousands of terrestrial years, in which the visible planets returned to their original sky positions. If they were aligned in conjunction, they'd return again to that position. The link between components and repetition stems from the fact that in a closed system (which the heavens were thought to be at the time) a finite number of elements can be arranged in only a finite number of ways before all of the possibilities are exhausted. Roll a set of dice enough times, and eventually the same combinations will repeat themselves. While most Greeks believed in cycles, they differed on what components were being recycled. What constituted the smallest things—which would add together in various permutations to form the stuff of Earth and the heavens—was the subject of much philosophical musing.

For example, the Pythagoreans—including the fifth-century BCE philosopher Philolaus, whose writings are preserved in fragments and quoted by later sources—believed that "all is number." The properties of odd and even numbers, particularly the first ten integers, governed the workings of the world and the sky in a fundamental way, they argued. Odd numbers represented light and good qualities; even numbers, darkness and evil.

Pythagorean cosmology is based on the simple geometry of circles, and the numerological associations of the number ten. Ten is sacred because it is the sum of the integers one through four—which can be arranged as dots in a triangle of four rows. Following circular paths, Earth, the sun, the moon, the starry dome, and the five visible planets—Mercury, Venus, Mars, Jupiter, and Saturn—orbit a great Central Fire (not the sun, but another source of power). That adds up to nine orbiting bodies, with Earth on par with the others. To complete the set of ten objects, as required by their numerological beliefs, the Pythagoreans posited a "counter-Earth" that could never be seen because it was always on the far side of the Central Fire.

Another ancient Greek philosopher, Empedocles, argued, in contrast, that nature's fundamental ingredients are air, water, earth, and fire. These are mixed together in various combinations through the attractive force of "love," and separated into purer forms by means of the repulsive agent, "strife." The actions of mixing and separation, in which love and strife alternately reign, result in grand cosmic cycles of creation and destruction.

Today we think of atoms constituting the most fundamental ingredients of ordinary matter, combining to form innumerable types of molecules—from the sodium and chlorine atoms in salt molecules, to the carbon, hydrogen, nitrogen, oxygen, and phosphorus in DNA. Before the modern concept, however, was an ancient Greek term: *atomos*, meaning indivisible. As the philosopher Democritus described such entities, if you cut anything into its smallest pieces, those would be its atoms. Atoms' main differences, he argued, are their shapes, which would differ between those making up sharp objects such as knives, smooth objects such as flowing water, and so forth.

Image 1. English mathematician and physicist Isaac Newton, who developed the laws of motion that govern classical physics. Credit: AIP Emilio Segre Visual Archives, Physics Today Collection.

It was more than two millennia later, in the seventeenth, eighteenth, and nineteenth centuries CE, when the scientific insights of thinkers such as Isaac Newton, Pierre-Simon Laplace, and John Dalton established the true nature of material things, including their dynamics and fundamental ingredients. Newton demonstrated in his "first law of motion" that, in contrast to the thinking of one of the most influential Greek philosophers, Aristotle, the natural state of objects is inertia: either remaining at rest or moving in a straight line at constant speed.

Aristotle had argued, wrongly, that any kind of motion, including constant velocity, required some kind of agent. Rather, as Newton showed, inertia does not involve forces. They are needed, in contrast, to break the state of inertia, and cause acceleration. Objects accelerate due to the action of forces, he asserted in his "second law of motion," with the amount of acceleration inversely proportional to those objects' masses. In other words, force equals mass times acceleration.

Making use of Newton's laws of motion, Laplace showed how by knowing the positions and velocities of a set of objects in space, as well as their mutual forces, one might predict their motions indefinitely forward in time. Presuming that everything is made of particles governed by Newtonian mechanics, that implies that complete knowledge of the state of the universe at any time (a tall order indeed!) offers the prospect of being able to forecast its entire future. Such a concept is called Laplacian determinism.

Borrowing Democritus's term, but altering the concept to explain testable chemical properties, Dalton developed the notion that compounds are made of atoms in various proportions, each corresponding to the smallest possible amount of a chemical element that still retains the element's identity. Atoms corresponding to different elements, such as hydrogen, carbon, and oxygen, have different masses. In arguing that all matter is made of atoms, he bolstered a conjecture by Newton that corpuscles, or particles, are the fundaments of nature. Dalton's atomism enhanced the concept of predictability in classical physics, by allowing for a means to imagine minute ingredients interacting with each other via well-defined forces, and thus predicting their future behavior.

In tandem, these ideas spurred a revival of the notion of repetition in space and time. After all, if, like checkers and chess, the cosmos has a finite set of pieces, and its moves are fully determined, chances are it is bound to repeat the same moves eventually. Earth's events, therefore, would be likely to recur elsewhere in space or sometime in the future. The uniqueness of our lives, in that case, would be an illusion.

BEYOND FORTRESS EARTH

French socialist leader Louis-Auguste Blanqui was a stubborn revolutionary. Like a comet in a closed orbit, his blazing actions only seemed to lead him back to the same place: prison. Throughout the mid-nineteenth century, he raged against the succession of authoritarian and bourgeois leaders who ruled in Paris, organizing worker demonstrations and armed bands of radicals in unsuccessful efforts to establish rule of the people. He ended up spending—in various jails and intervals of incarceration—more than thirty-three years of his life imprisoned, earning him the nickname "the locked-up one."

Perhaps it is an emblem of the human condition, that in one of his strictest confinements—the island fortress of Taureau in the English Channel—Blanqui's mind soared toward the stars, seeking intellectual freedom. It was a bleak time in his life, shortly after the defeat of the Paris Commune, when thousands of his former comrades in arms had been locked up or killed—with Blanqui absent from the scene due to a previous prison sentence. In 1872, while housed in his stark, frigid prison cell on that windswept isle, he put aside politics, momentarily, for the pursuit of science and wrote the imaginative treatise *Eternity Through the Stars*. Arguably, though it confines itself to a single universe, by contemplating an endless number of worlds in space, some identical (or nearly identical) to Earth, the work is an important intellectual antecedent of the modern concept of a multiverse. In it, an imprisoned soul offered one of the quintessential guides to humanity's place in an immeasurably vast, and potentially repeating, cosmos.

Blanqui based his thesis about replica Earths on several key propositions. One is that the universe is infinite and relatively uniform, harboring an endless succession of stars, including many similar to the sun. Moreover, the bulk of those stars are orbited by planets. Hence countless planetary systems speckle space in all directions.

The notions of infinite worlds and the non-centrality of Earth were scarcely novel ideas at the time of Blanqui's treatise. The latter dates at least as far back as the Pythagorean proposition, expressed by Philolaus, that Earth and the other visible planets orbit the Central Fire in similar fashion. The implication was that Earth is not particularly special—Jupiter, Saturn, and so forth would each have equal claim. Centuries later, Aristarchus of Samos (310–230 BCE) modified that system by banishing the unknown Central Fire and making it heliocentric (the sun at the center). Earth, he rightly argued, revolves around the sun and rotates on an axis, leading to the progressions of day and night, as well as the seasons.

That said, a geocentric (Earth at the center) belief took hold in much of Europe and North Africa, following the advocacy of Aristotle (who lived before Aristarchus, but whose authority greatly outlasted his lifetime) and others that the sun, moon, and visible planets orbit the Earth. Especially after Ptolemy's influential Almagest, written about 150 CE, described an intricate system of cycles within cycles, involving the sun, moon, planets, and other celestial bodies, centered on Earth, to explain what is seen in the nocturnal sky, geocentric astronomy became canonical for centuries, particularly in Christian thought. Nicely for Christian theologians, it well matched biblical descriptions of the sun rising in the sky, and it corresponded suitably with the notion that terrestrial religious events, such as the story of Adam and Eve as the first humans, were unique and central. By the time of the Middle Ages in Europe, doubting the geocentrism of Aristotle and Ptolemy was tantamount to heresy.

During the Renaissance, however, scholarship began to broaden, and the wide gamut of Greek thinking was rediscovered. Dissident voices began to argue again for the relative simplicity of a heliocentric system, and also for the non-uniqueness of Earth. Most notably, in 1543 CE, Nicolaus Copernicus of Toruń, Poland, published *On the*

Revolutions of the Heavenly Orbs, a brilliant revival of Aristarchus's hypothesis, but far more fleshed out by matching astronomical data known at the time. In particular, it took data collected by Ptolemy using naked-eye astronomy, and matched it to a model in which the sun is at the center of a solar system of concentric planetary orbits. Copernicus couldn't quite match the data with pure circular orbits, so similarly to Ptolemy he added epicycles—small circles around which planets turned as they navigated the larger circles around the sun.

Many decades later, based on the superior data collected by the last great naked-eye astronomer, Tycho Brahe, the sixteenth-to-seventeenth-century German mathematician Johannes Kepler greatly improved the Copernican system by eliminating the epicycles, and replacing the circular orbits with ellipses. In Kepler's model, each planet, including Earth, follows an elliptical orbit, with the sun as one of the focal points (the two focal points of an ellipse constitute the generalization of the center of a circle). Moreover, he showed that the time taken for a planet to complete its orbit depends on its distance from the sun.

In pondering the notion of infinite worlds, Blanqui walked in the footsteps of yet another Renaissance thinker, Giordano Bruno, a near-contemporary of Kepler's. Bruno's 1584 treatise, *On the Infinite Universe and Worlds*, criticized many of Aristotle's ideas about nature, including his geocentrism, and argued that Earth was far from unique. Extending the Copernican notion that Earth orbits the sun to other stars, Bruno argued that space, and its number of planetary systems, were endless. Many of Bruno's views on theology were similarly antithetical to Church teachings. He traveled around Europe for a number of years preaching his views freely, but made the mistake of returning to his native Italy. There, he was denounced to the Inquisition. Tried in Rome, he was sentenced to death and burned at the stake in 1600.

Blanqui's own heresies were political, not theological—he opposed the concentration of wealth in the hands of French aristocrats and the rich, in general. By his time, thanks to the telescopic observations of Galileo and his successors, and the works of Newton and others, the notion of Earth orbiting the sun, which was itself one of numerous stars, was widely accepted. Therefore, in and of itself, Blanqui's advocacy of an infinite universe with an infinite number of planets was not so radical for his time.

What was particularly innovative was adding atomism and the laws of chance to the mix. Blanqui argued that in an unlimited cosmos with an endless array of planets, but a finite number of elements, repetition and near-repetition are bound to occur. Earth's turbulent history, after all, is based on the interaction of people and things made of various chemicals. As Russian chemist Dmitri Mendeleev had shown elegantly in 1869, the chemical elements might neatly be arranged, according to their properties, in a periodic table. Such an arrangement lent considerable weight to Dalton's concept of atomism and the notion that everything on Earth—and indeed likely in the universe—is based on arrangements of atoms as building blocks. Around the same time as Mendeleev's classification, William Huggins, later joined by Margaret Huggins as a husband-wife team, aimed spectroscopes (instruments that divide light into its individual spectral lines, much like a rainbow of color) at an array of astral bodies, such as stars and nebulas (gaseous clouds, or, in some cases, misidentified galaxies), showing they had spectral lines matching those of terrestrial elements such as hydrogen and nitrogen. Thus, by the time of Blanqui, the notion that countless planets are made of the same atoms as Earth, more or less, but in different combinations, had strong support.

Imagine the atomic makeup of a planet as a colossal deck of cards, shuffled in myriad ways to display an enormous, but finite, number of arrangements. In an infinite space, the same or similar arrays are bound to happen again and again. Thus, Blanqui envisioned near-identical copies of Earth, with key differences such as Napoleon defeating the Duke of Wellington in the Battle of Waterloo, instead of losing to him. On the other hand, Blanqui thought, there would necessarily be an unlimited number of exact replica Earths in space, for which those who are cursed with tragedy on this world, such as he was, would be doomed on all the other versions too. He found that idea of parallel worlds of identical tragedies to be a singularly depressing thought.

THE SNARE OF ETERNAL RETURN

Merely a decade after Blanqui completed his treatise, German philosopher Friedrich Nietzsche arrived at a similar notion of replication of history, involving time rather than space. Nietzsche dubbed the concept of endless repetition of events throughout the eons "eternal return," also known as "eternal recurrence."

In contrast to the notions of similar, but not identical, cycles pervading many ancient cultures around the world, including, for example, the Hindus, Maya, and Greeks, Nietzsche focused on the concept of exact repetition over time. Putting aside the idea of decay, prevalent at the time because of the laws of thermodynamics showing how disorder tends to grow, he concentrated on the microscopic properties of the materials that make up people and their environments. Adopting the materialist viewpoint that even our thoughts and sense of free will are deeply determined by our underlying physical mechanisms, he concluded that, given enough time, the arrangements needed for particular thoughts, feelings, and events are bound to happen again and again. Replica versions of him and everyone else would relive their lives precisely, like actors in a nightly theatrical performance, the majority completely unaware that they've already enacted the same randomly penned "script" countless times.

Nietzsche called his "discovery" of eternal return a sudden revelation, but he may have encountered cyclic ideas in his readings—

conceivably even learning about Blanqui's idea. Like Blanqui, Nietzsche led a life of misery. He was sickly, and unlucky in love. For him, the notion that chance recombination of atoms could bring recurrence after many thousands, millions, or billions of years, seemed at first to him the greatest horror.

As Nietzsche wrote in his 1882 treatise *Die fröhliche Wissenschaft* (*The Joyful Science*):

> What if some day or night a demon were to steal after you into your loneliest loneliness and say to you: "This life as you now live it and have lived it, you will have to live once more and innumerable times more; and there will be nothing new in it, but every pain and every joy and every thought and sigh . . ." How well disposed would you have to become to . . . crave nothing more fervently than this ultimate eternal confirmation and seal?

Then, as he contemplated eternal return further, he began to revel in the idea that he was the discoverer (at least supposedly) and leading advocate of the notion. That meant that in future incarnations he would similarly be its prophet. Consequently, his place in history, he thought, would be sealed forever. Grandiosity began to alternate with his melancholy.

In his later years, Nietzsche suffered from debilitating mental illness, likely due to a dire physiological condition of some kind. He stepped down from all positions of responsibility, stopped publishing his works, and spent much of his time wandering by himself in natural settings, such as forests and mountain paths. Meanwhile, his health continued to deteriorate. After several strokes, he died in 1900.

In death, though, Nietzsche's voice was not silent. Posthumously, writings of his emerged revealing that he had continued to develop the concept of eternal return even during his isolated years. He wanted his legacy to be cemented in solid scientific reasoning, not just conjecture based on feelings. To bolster his argument, his later work centered on the idea of using conservation of energy, a principle derived from Newtonian classical mechanics, to justify the notion of recurrence.

ENDLESS LOOPS: CONSERVATION LAWS AND SYMMETRY

Newton's laws of motion are brilliant in their simplicity and power. As Laplace and others emphasized, they offer a way to use the positions, velocities, and forces of all objects at any given time in the history of the universe, to calculate the next step for those bodies, the step beyond that, and so forth, indefinitely into the future. Even though we now know some of the limitations of Newton's mechanical methods, we trust them enough for many complex calculations about the trajectories of spacecraft—such as taking astronauts multiple times to the moon and back.

To truly exploit the implications of Newton's principles, however, requires a handy method called conservation laws. A conservation law involves finding a physical parameter or combination of parameters that demonstrably remains constant over time. One key example is linear momentum, a measure of the tendency for the central point—specifically a weighted average, called the "center of mass"—of an object or set of objects to move at the same velocity in a straight line, unless an external force influences that object. Picture a pair of Olympic skaters gliding effortlessly forward, hand-in-hand, in an icy rink, and that is linear momentum conservation. Even if one of the skaters let go, the center of mass of the pair would continue to move forward at the same rate, and the motion of the other skater would adjust accordingly to compensate (if, in letting go, one veered to the left, the other would veer to the right for balance).

As brilliant German mathematician Emmy Noether would rigorously demonstrate in the 1910s, each conservation law is associated with a symmetry of nature. For example, conservation of linear momentum is connected with "translational symmetry": the idea that if you slide the object in question along a straight line, a process called "translation," nothing in its environment changes. Of course, that is an idealization for a skating rink—there might be tiny bumps or irregularities in the ice—but its essence is true: skating rinks strive to be flat and smooth.

Another important natural symmetry is "rotational symmetry": the uniformity associated with spinning a completely circular wheel, wholly free of impediments and other external influences. Imagine a bicycle's tires effortlessly turning as they roll along a country lane. The conservation law matched with that symmetry is conservation of angular momentum. It is so powerful that it helps keep the bicycle upright when it is moving. Whirling bicycle wheels resist toppling over because it would involve a change in angular momentum. Maintaining constancy of angular momentum thereby preserves stability.

Finally, let's consider "time symmetry": the ageless attributes of a system. For instance, picture an ideal, frictionless grandfather clock with a pendulum that swings back and forth indefinitely. Of course, real clocks need to be wound, but let's put that aside for now. If, in a bout of insomnia, you stay up all night watching the pendulum swinging, it would look exactly the same at 2:00 a.m., 3:00 a.m., and 4:00 a.m. The conservation principle linked to that is the "conservation of mechanical energy."

Mechanical energy can be defined as the sum of "potential energy," the energy of position, and "kinetic energy," the energy of motion. Each is constructed from Newton's laws using calculus. For certain types of ideal systems—without damping (motion-reducing) forces, such as friction and air resistance—mechanical energy remains constant over time. That is, potential energy and kinetic energy might transform from one into another—like an hourglass releasing sand from its upper to lower halves—but the total, mechanical energy, stays the same.

Potential energy is an attribute that applies only to certain types of forces, called "conservative forces." Conservative forces speed up objects in a manner that depends on their initial positions relative to a baseline—the farther away, the more potential they have to increase their speed. A good example is gravitation, for which height above a certain level, such as the ground, determines an object's prospects for motion. Thus, gravitational potential energy depends on height.

Drop a ball from a certain height above the floor, or place it on a frictionless slope, starting at the same height, and either situation would correspond to the same gravitational potential energy. Watch it fall or slide, and, in either case, its speed just before it hits the floor would be the same. That ultimate speed, along with the ball's mass, is a factor in its final kinetic energy. Hence, during the process of falling or sliding, potential energy, due to height, converts to kinetic energy. At the top, the potential energy is maximum and the kinetic energy is minimum. Near the floor, the potential energy is minimum and the kinetic energy is maximum. The total mechanical energy—potential plus kinetic—remains the same: only the type changes, in a process of recycling.

In the case of the pendulum of a grandfather clock, each time it swings to its highest position, its potential energy is maximum. As it lowers, that energy would convert into kinetic energy. At the bottom, its kinetic energy would reach its maximum value, and its speed would be greatest. Then, it would begin to rise again back to its original height, turning the kinetic energy back into potential. Assuming, once again, the ideal case of no damping forces, the pendulum's cycles would repeat again and again, making the clock tick at a uniform pace.

Note that any kind of damping forces, such as friction or air resistance, would act to diminish the overall mechanical energy, by transforming some of it into heat. In that case the symmetry would be broken—later times would look different than earlier times—and mechanical energy would not be conserved. The lesson is that broken symmetries lead to shattered conservation laws. Keep in mind, though, that a more general form of the conservation of energy, which includes heat, electrical energy, nuclear energy, and so forth, along with mechanical energy, would still be maintained.

Conservation laws offer a critical way of making predictions without the need for direct observation. For instance, imagine a frictionless roller coaster being released from the top of a lift hill, sliding down, and passing through a dark tunnel. Presuming mechanical energy is conserved, even without seeing it as it zooms through the darkness, we'd readily be able to predict the time when it would leave the tunnel, and how fast it would be moving.

For celestial bodies, conservation laws enable us to anticipate their cyclic behavior. For instance, the steady orbits of planets and satellites, such as Mars around the sun and the moon around Earth, can be explained and prognosticated by means of conservation of angular momentum and conservation of energy.

In his popular book *Fragments of Science*, published in 1871, Irish physicist John Tyndall exalted in the predictive power of the law of energy conservation. He saw it as a tool by which all of nature might be understood through Newtonian physics, including the mysteries of life itself. The ironclad progression of cause and effect would rule supreme. As he wrote:

> The doctrine of the Conservation of Energy, the ultimate philosophical issues of which are as yet but dimly seen, . . . "binds nature fast in fate," to an extent not hitherto recognised, exacting from every antecedent its equivalent consequent, from every consequent its equivalent antecedent, and bringing vital as well as physical phenomena under the dominion of that law of causal connection which, so far as the human understanding has yet pierced, asserts itself everywhere in nature.[1]

Tyndall's logic, applied to a finite space containing a finite number of ingredients run for an unlimited time, mandates that patterns are bound eventually to repeat themselves. Determinism ensures that once a set of particles, after a lengthy interval, happens to return to the same positions, velocities, and other conditions due to energy conservation, it would reproduce the same behavior. Nietzsche, in his final years, used similar reasoning to justify his notion of eternal return. As if riding on an endless, frictionless roller coaster, humankind would be doomed to repeat over and over the same dizzying highs and gut-wrenching drops.

THE SLOWEST CLOCK: TIMING POINCARÉ RECURRENCE

The late nineteenth century saw great leaps forward in applying mathematical analysis, including graphing, to the study of dynamics. Dynamics is the physics of how forces affect motion. Powerful methods emerged that are still used today. To analyze the dynamics of systems with finite elements, theorists often plot the behavior of each particle along separate axes, in a type of graph with multiple axes called "phase space." Some of the axes might represent each particle's position coordinates, such as x, y, and z. The remaining axes might represent each particle's velocity or linear momentum (mass times velocity) components. By tracking the position and velocity of all particles over time, one might identify any predictable behavior of the system, such as periodicity.

Trying to picture such a graph might be challenging, however, because of the sheer number of axes. If there are six axes for each particle, three components of position and three components of velocity, and multiple particles, that quickly adds up to numerous mathematical dimensions. Rather than an actual physical space, such mathematical dimensions span an abstract space of information that happens to have multiple coordinate axes. For instance, weather forecasters plotting the temperature, pressure, wind speed, and height of various atmospheric readings, using a four-dimensional graph, would be representing that information within the context of a higher-dimensional abstract space, not an actual space. They'd be employing mathematical, rather than tangible, dimensions.

The work of brilliant nineteenth-century thinkers Carl Gauss, Bernhard Riemann, and others produced an explosion of interest in mathematical dimensions. Higher dimensions became seen as tools to understand the full range of algebraic, geometric, and topological properties the realm of mathematics has to offer. While nature's craftsmanship appeared limited to three dimensions (we now add time as the fourth), abstract mathematics seemed adept at sculpting higher-dimensional creations. For example, if line segments have two endpoints, squares have four linear edges, and cubes have six square faces, the obvious generalization is the hypercube, the four-dimensional analogy of a cube, with eight cubic sides. Indeed, in the esoteric realm of pure geometry, one might consider an endless number of dimensions. But, accustomed to the limited vistas of length, width, and height, how might our minds picture them?

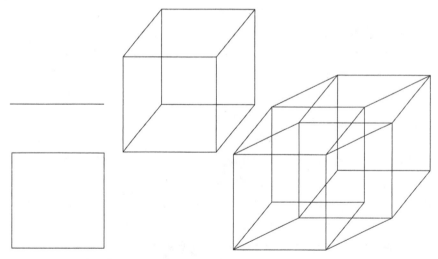

Image 2. Regular shapes in various numbers of dimensions. From top to bottom: A one-dimensional line segment has two ends. A two-dimensional square possesses four edges. Six square faces border a three-dimensional cube. Extrapolating into four dimensions, one might envision a hypercube, also known as a tesseract, as being composed of eight connected cubes.

In that visionary era, several authors of fiction sought to introduce the concept of dimensionality to the general public. In 1884, City of London School headmaster Edwin A. Abbott published his marvelously imaginative novel *Flatland*, which envisioned a two-dimensional society with geometric beings, such as squares and circles, residing in a plane. Several years later, mathematician Charles Hinton coined the term "tesseract," another word for four-dimensional hypercube, and emphasized how it might be envisioned by means of its shadow in three dimensions. Abbott and Hinton, in their respective works, urged the public to be open-minded about the possibility of kingdoms beyond ordinary space.

One of the most versatile thinkers of the day was French mathematician Henri Poincaré. Poincaré's interests spanned the practical and abstruse, from serving as a national inspector investigating safety conditions in mines to proposing a conjecture about spheres and hyperspheres (four-dimensional generalizations of spheres) in the field of geometric topology that took almost a century for other mathematicians to prove.

In 1890, Poincaré published a remarkable treatise, "On the Three-Body Problem and the Equations of Dynamics," that, at first glance, seemed to support the eternal return hypothesis. Analyzing a system of many particles obeying the conservation of energy, he showed that if one began with an arbitrary state and waited long enough, that initial state would be approached (very nearly repeated) again and again endlessly. For any energy-conserving system, therefore, one might always calculate a "Poincaré recurrence time," in which its dynamics essentially repeats itself. Given that people's actions depend on the behavior of the atoms and molecules in their bodies, that would seem to imply that, given a long-enough wait, history would eventually repeat itself.

Poincaré's result surprised many physicists because it seemed to defy the widely supported Second Law of Thermodynamics, introduced by Rudolf Clausius in the 1850s, stating that in a closed system, entropy, a measure of unusable energy, tends to remain the same or increase, not decrease. Systems with distinctive components, such as ice cubes floating in a steaming pot of hot water, tend to have lower entropy overall than uniformly mixed systems, such as tepid water in a glass. On the molecular level, the former would represent groupings of slower-moving water molecules (the cubes) in contact with a section of faster-moving molecules (the hot water), while the latter would comprise molecules of intermediate speeds. The Second Law mandates that if ice in a pot of hot water is left on its own it might increase its entropy by transforming into a lukewarm liquid, but never the opposite. That seems like common sense—we never see ice naturally emerging from room-temperature water and leaving the rest of it hot, without a rigged contraption, such as a combination of freezing and heating coils, to carry out the process. At any rate, such a contraption would increase entropy in its own right. However, according to the notion of Poincaré recurrence, wait long enough, and the system's dynamics would eventually bring it back to a near-replica of the original molecular distribution—namely ice cubes floating in hot water. In seeming contradiction to the Second Law, entropy would spontaneously decrease. How could that be?

Image 3. French mathematician Henri Poincaré, who innovated the study of dynamics. Credit: The University of Chicago Yerkes Observatory, courtesy of AIP Emilio Segrè Visual Archives, Physics Today Collection and Tenn Collection.

Before tossing aside the Second Law and anointing Nietzsche the prophet of eternal return (as he undoubtedly would have wished to be remembered), we must now look at the fine print. Poincaré's paper predicts that the greater the number of a system's components, the longer its recurrence time, by an increasingly greater amount. For a system of a few particles, the recurrence time would likely be in the fractions of a second, but for a tall glass of water with some 10^{25} (1 followed by 25 zeroes) molecules, the recurrence time would be astronomically large. Any return to its original state would essentially be impossible, as it would take far, far longer than the age of the universe. That is, you could wait until the sun and all the other stars die before a glass of tepid water would spontaneously rearrange its molecules into slower and faster segments to produce ice cubes along with boiling water.

And that's just a glass of water. The time for all of the molecules on Earth to completely revert to any given state would be far, far longer than that, making all the time since the Big Bang seem like a flash.

Nietzsche's vision of history recurring because of conservation laws, therefore, could never realistically happen within the universe's timeline—it would simply take far too long. If the universe is infinite, on the other hand, it is possible that chance assembly could create a near-replica Earth, fulfilling Blanqui's vision. But that world would undoubtedly be so far away that it would lie well beyond observability.

Given how long it would take for terrestrial events to repeat themselves, trying to calculate the recurrence time of the observable universe would seem to be an exercise in futility. Nevertheless, it's mind-boggling to try to grapple with that unimaginably immense figure, estimated in 1994 by physicist Don Page to be on the order of $10^{\wedge} 10^{\wedge} 10^{\wedge} 100$ (ten raised to the power of ten raised to the power of ten raised to the power of one hundred) years.[2] Absolutely inconceivable!

Still, boosters of Newtonian methods could point to its predictive power for basic systems with only a few components. Those assertions would soon be tempered, however, by the emergence of what came to be known as chaos theory: the sensitivity of certain deterministic systems to their initial conditions. Poincaré would point out that issue himself. For some types of dynamics, if you get the initial conditions wrong by the tiniest amount, that error would blow up over time, rendering nil your chances of anticipating its future state. Practical considerations, such as imprecise data, sometimes muddy-over Newtonian physics' otherwise clean window into the future.

MAXWELL'S FIELDS: THE TUMBLING OF NEWTON'S APPLE

In the final decades of the nineteenth century, physicists began to realize that while Newtonian mechanics possesses considerable predictive strengths (aside from the issue of sensitivity to initial conditions identified later in chaos theory), it also has glaring conceptual weaknesses. Fundamentally, concepts as basic as force and inertia are tricky to define, as Newton himself recognized.

Newton imagined long-range forces, such as gravitation, as being invisible "threads" drawing together objects, sometimes over immense distances. Such interacting things, he realized, ranged from terrestrial to astronomical. Reportedly, his venture into gravitational theory began with observing an apple falling from a tree, and later generalizing the notion to other objects that have mass. From the apple to the moon, each would have a distant, unseen attraction to Earth or another massive body.

As reported by Newton's friend, archeologist and biographer William Stukeley, referring to a late-life discussion between the two of them in a garden setting:

> After dinner, the weather being warm, we went into the garden and drank tea, under the shade of some apple trees. . . . He

told me, he was just in the same situation, as when formerly, the notion of gravitation came into his mind. It was occasion'd by the fall of an apple, as he sat in contemplative mood. Why should that apple always descend perpendicularly to the ground, thought he to himself. . . .[3]

The conceptual issue regarding Newton's description of gravitation and other forces involves the intermediate space between the interacting bodies. For an apple falling from a tree down to the ground that space isn't far, but what about the considerable distances between the sun and its orbiting planets? How do Saturn and Jupiter, for instance, sense that the sun is out there—hundreds of millions of miles away—and feel its pull? If the sun suddenly disappeared, how would they know? Would the changes in their motions due to the sun's vanishing be instant, or would they take time? Newton's concept of gravitation and other forces failed to address such questions.

Another conceptual quandary pertaining to Newton's laws involves the concept of inertia as the natural state in the absence of forces (or if all of the forces on a body are precisely balanced). Inertia involves either moving at constant velocity, or not moving at all. In either case, the question arises: motion or lack of motion relative to what? At a loss to explain inertia otherwise, Newton developed the notion of "absolute space" and "absolute time," as hypothetical yardsticks and clocks in space that serve as reference frames. Given that everything in space, from comets to stars, is constantly in motion, it became unclear to scientists where such reference frames were. In particular, the late nineteenth-century Austrian philosopher Ernst Mach savaged the concept of absolute space and time as intangible, suggesting modifying the concept of inertia to be an effect of the influence of the distant stars themselves.

The 1800s witnessed great advances in the study of electricity and magnetism, culminating in brilliant Scottish physicist James Clerk

Maxwell's united theory of electromagnetism. Drawing on the work of Gauss, Michael Faraday, André-Marie Ampère, and others, Maxwell described electromagnetism as a local field theory, rather than in terms of forces acting over a distance. The concept of fields as conveyors of forces through space was a much-needed update of Newton's work.

A field is an entity that has a value, or set of values, at every point in space. Typically, if it involves a single value at each point, it is called a "scalar field." For multiple values at each point, it is called a "vector field" or "tensor field," depending on certain mathematical properties. If the coordinate system is rotated, vectors transform in a simpler way than do tensors, and scalars remain the same.

Mathematical terminology can sometimes cause stress for those prone to anxiety. To alleviate that, imagine the soothing properties of a heated, swirling whirlpool. Now back to the math. If thermometers measure the temperature of the pool at each point, those values constitute a scalar field. There's a single temperature value matched to each location in the water. Even if you rotate your perspective, by viewing the pool from the opposite direction the map of temperatures would stay the same, fitting for a scalar field. Now, picture flowmeters at each spot, recording the speed and angular direction of the water. A map of those velocities, featuring multiple components—magnitude and direction—at each point would comprise a vector field. Moreover, if you rotated your perspective, those angles would change, but in a simple, uniform way—another characteristic of vector fields. Finally, to generate a tensor field, imagine pressing on the water in each direction everywhere in the pool, and seeing exactly how it responds in each direction. For example, imagine recording how much the water would compress or expand in the left direction if being pressed downward. If there are three directions for pressing, and three directions for reacting, that corresponds to three times three components at each position. Voilà, a nine-component tensor field. Clearly, rotating one's perspective would transform those components in a more complicated way than if it were a vector field.

By the time of Maxwell, physicists long realized that vector fields offered the perfect way to describe either electricity or magnetism. Take a bar magnet, place it on a flat sheet of paper, and sprinkle iron filings throughout the sheet. Amazingly, the filings will typically start to arrange themselves along distinct curves, branching out from the magnet's poles. Trace each of those curves from the north pole of the magnet to the south pole, and those are the magnetic field lines. The strength of the field at each point matches the closeness of the field lines, and the direction matches their angles. Clearly, therefore, magnetism is well-represented by a vector field. Similarly, electricity near charged objects can be described by a vector field as well.

Force comes into play, not via mysterious "threads" connecting distant objects, but rather from direct exposure to electric and magnetic fields. Place a moving charge in any combination of electric and magnetic fields, and it would experience electric and magnetic forces, respectively. Generally, these forces would be perpendicular to each other. The effect would be local, based solely on conditions in the charge's environment, rather than any remote action.

Following numerous experiments showing how electricity and magnetism are connected—for instance, electric coils create magnetic fields—Maxwell decided to describe both as a combined electromagnetic field with perpendicular electromagnetic components. He recognized that, like tossing a pebble into a lake and creating ripples, oscillating charges would trigger electromagnetic waves. That realization led him to combine all of the different equations describing how electric and magnetic fields are created into a single wave equation characterizing how electromagnetic fields propagate through space. Calculating the speed of such electromagnetic waves, he discovered that it matched the velocity of light. Consequently, he arrived at the astonishing conclusion that electromagnetic radiation is light, in its various visible or invisible forms.

Today we speak of the vacuum speed of light, with "vacuum" meaning as close to sheer emptiness as we can imagine. At the time of Maxwell's electromagnetic wave proposal, however, the mainstream scientific community was attached to the view that all waves must oscillate through some kind of material. Ocean waves travel through water. Wind gusts and breezes travel through air. Earthquake tremors travel through the ground. For light as an electromagnetic wave, the hypothetical substance through which it traveled was called the "luminiferous ether," or ether (sometimes spelled "aether") for short.

If space were filled with ether, Earth would experience an "ether wind" rushing by it as it moves. Just as the speed of a sailboat depends on its movement relative to air and water currents, logically, the motion of light as Earth moves against the ether wind should be distinguishable from motion in the wind's direction. On the contrary, numerous experiments designed to find differences in the speed of light due to the ether wind have found absolutely none. The most famous and influential of such measurements was an interferometry (precision timing of light waves) experiment conducted in 1887 by Polish-American physicist Albert Michelson and American chemist Edward Morley, that found no distinction in light's speed if it is traveling in two different perpendicular directions. The expectation was that the ether wind, if it existed, would make a difference, but absolutely none was found. That negative result, which would be reproduced many times, would eventually prove the death knell for the concept of luminiferous ether—although it took many decades for the scientific community to discard the idea completely.

Maxwell's theory gave a constant value for the speed of light that all observers would measure no matter their circumstances. The Michelson-Morley experiment and other precision measurements each supported that view. The idea of light having a uniform speed in empty space, regardless of the velocity of the observer, stood in blatant contrast, however, to a long-standing notion called "Galilean relativity." Named after an observation by Galileo about the relative velocities of moving objects being different from their absolute velocities (compared to a fixed reference frame), the concept was embodied in Newton's laws of motion. If two different passengers were standing next to each other on twin parallel moving walkways (in an airport terminal, for instance) that exactly kept pace with each other, they'd seem to each other to be standing still. Applying that principle to light however, led to a contradiction. Galilean relativity predicts that someone on an ultra-high-speed moving walkway, somehow keeping pace with a beam of light, would see it as still. Maxwell's theory, on the other hand, demands that light's speed remain constant no matter the velocity of the observer. In 1894, at the age of fifteen, young Albert Einstein would imagine trying to chase a light wave, and begin searching for a solution to that dilemma—which would culminate, more than a decade later, in his special theory of relativity.

In the meanwhile, despite the Michelson-Morley experiment, belief in an invisible ether persisted. In certain circles, discussions of ethereal currents blended with talk of unseen higher dimensions in explorations of the meaning of consciousness and the realm of the spirit. Well before multiverse notions emerged, the scientific community was divided about the role in natural inquiry of domains beyond the observable.

THE UNSEEN UNIVERSE: ETHER, SPIRIT, AND HYPERSPACE

In 1874, Tyndall delivered a speech in Belfast to the British Association for the Advancement of Science, making a passionate case for belief in objective, experimental science, and rejecting the unobservable. As in *Fragments of Science*, he extolled the importance of atomism and the principle of conservation of energy, leading to extraordinary powers of scientific prediction using Newtonian physics and other means, such as Darwinism. In contrast to such scientific triumphs, Tyndall dismissed supernatural beliefs as relics of the past. Religion's place, he argued, was to offer emotional support and moral guidance, but had absolutely nothing to do with objective reality. Cosmology, the study of the universe, must become exclusively the enclave of science, and eschew connections with religion.

As Tyndall remarked: "The impregnable position of science may be described in a few words. We claim, and we shall wrest, from theology the entire domain of cosmological theory. All schemes and systems which thus infringe upon the domain of science must, *in so far as they do this*, submit to its control, and relinquish all thought of controlling it."[4]

The following year, in rebuttal to Tyndall's speech, two noted Scottish physicists, Balfour Stewart and Peter Tait, published *The Unseen Universe*. Stewart and Tait saw Tyndall as part of a growing minority of thinkers erroneously arguing for strict materialism to the exclusion of religion. In contrast to such materialists, the authors argued for scientific exploration of the realm of the spirit, including the mysteries of the soul and the afterlife—what we would now call the paranormal. For example, if conservation laws could be applied to the soul, they reasoned, it would represent an immortal substance, matching what religious thinkers have asserted.

Stewart was a prominent example of British, German, and other European late nineteenth-century scientists who began to explore in earnest all manner of alleged psychic phenomena, from so-called mediums claiming to contact deceased relatives of the bereaved during séances to those purporting to read minds. Others similarly interested in the paranormal, at least to some extent, included British chemist William Crookes, British naturalist Alfred Russel Wallace, and German physicist Wilhelm Weber. Sherlock Holmes writer Arthur Conan Doyle also became a devotee.

Stewart's interest in ether and the continuity of the soul as well as material things meshed well with his paranormal interests. From 1885 to 1887 he served as the second president of the London-based Society for Psychical Research (an investigatory group that has persisted to the present day).

Another accomplished late nineteenth-century scientist who became deeply interested in purported psychic phenomena was the German physicist Johann Zöllner, who was an expert in astrophysics and optical illusions. His investigative skills for detecting illusions fell short, however, when, starting in 1877, he was deceived by American self-proclaimed psychic medium Henry Slade. Slade had developed a knack for "slate writing," in which during a darkened séance—with everyone in a group sitting around a table holding hands on top of it, including him—a message would mysteriously appear on a slate placed underneath it. In reality, he would prepare the back of that slate, or another slate, beforehand, or else write with his toes clutching a piece of chalk—as debunkers of his methods would reveal. By the time he caught Zöllner's interest, he had already been found guilty of fraud at a well-publicized trial in London.

Zöllner's fascination with Slade derived mainly from his hope that the American medium's feats could shed light on the nature of the spatial fourth dimension. Talk of four and higher dimensions had become popular among mathematicians. In mathematics, the number of spatial dimensions appeared to have no limit. Hyperspace, including hypercubes, hyperspheres, and other generalizations of three-dimensional objects, seemed as geometrically realistic as cubes, spheres, and so forth. In physics, abstract phase spaces of numerous dimensions proved incredibly useful, as mentioned, in the study of dynamics. Why then, many wondered, did nature retire from its craft after introducing length, width, and height? Or did it furtively create higher dimensions? Some thinkers had already begun to speculate that time was the fourth dimension, due to the commonness of equations that possessed both spatial and temporary coordinates. But why not higher spatial dimensions as well?

Watching Slade perform several feats, such as locking together two solid wooden rings, tying knots on a rope held tightly on each end, and removing coins from a sealed box, Zöllner became convinced that he must somehow have access to a fourth spatial dimension, perpendicular to the others, and undetectable to the senses of most people. His gift, Zöllner thought, revealed that it is possible to twist, untwist, insert, and remove items using higher-dimensional manipulation. Otherwise, how could Slade's feats be possible? Of course, it was just the trickery of an adept magician, but Zöllner was fooled. With the zeal of a devotee, he penned a book, *Transcendental Physics*, to advance his hypothesis that access to the fourth dimension was possible.

Rather than convince the public of the reality of the fourth dimension, however, Zöllner's advocacy (in the context of Slade), along with promotion of higher-dimensional notions by members of spiritualist groups such as the Theosophists, served to associate higher dimensions with mysticism. Many mainstream scientists steered clear of the notion altogether. Zöllner died in 1882. Soon thereafter, his reputation began to wither due to the Seybert Commission, an investigation of spiritualism, focusing on Slade's work, funded by an endowment to the University of Pennsylvania. The experts on the panel concluded in their 1885 report that Slade employed noticeable trickery, which should have been apparent to reputable scientists such as Zöllner. The fact that Zöllner missed obvious clues reflected poorly on his state of mind, and perhaps indicated deterioration of his mental health, the report indicated.

Maxwell died in 1879, having witnessed only the start of this debate. Though a good friend of Tait's, and knowing others interested in spiritualism, he had stuck to serious science and avoided the question of the paranormal. Nor did he address the fourth dimension, except for a brief mention in "A Paradoxical Ode," a humorous poetic tribute to Tait.

Anticipating the early twenty-first-century debate between those who deem multiverse ideas "unscientific," and others who see them as ways to organize what we do observe into a more robust description, the late 1800s saw a rift between those who focused on the tangible and those willing to entertain the ethereal—in the broadest sense of the term. The dividing line wasn't always obvious, and the ultimate winner not always clear. For instance, Mach, who emphasized the evidence of the senses, criticized his colleague Ludwig Boltzmann's advocacy of atoms. There was no direct physical evidence for atoms at the time, so Mach suggested that they shouldn't be part of any theory until they could actually be sensed. Similarly, arguments between supporters and detractors of the notions of ether and of the fourth dimension centered on whether logical necessity was enough justification—motion needs a medium; hypercubes follow naturally from cubes—or if empirical proof was required.

As the nineteenth century drew to a close, the questions of the meaning of ether and of possible higher dimensions remained unresolved. Newtonian classical mechanics, particularly its reliance on Galilean relativity and absolute space and time, appeared to clash with Maxwell's conclusions about electromagnetism and the constancy of the speed of light. Both theories seemed experimentally sound, however. Dropping Newtonian physics, with its seeming ability to predict the future through deterministic methods, was not on the radar of the mainstream scientific community. Yet the field theory of Maxwell dazzled with its predictive power as well, plus it seemed more tangible with its analogy to fluid flow. Judging from what was written at the time, creating a self-consistent physics that encompassed all manner of natural phenomenon seemed tantalizingly close—and appeared to call for some creative thinking. How much weight should be given to unobservable elements, such as the purported, but undiscernible, ether, and the mathematical suggestion of a fourth dimension, lingered as unanswered issues. Today, contending factions of the physics community put forth strikingly similar arguments regarding the validity of various multiverse models. Though not exactly as Blanqui or Nietzsche imagined, history does indeed repeat itself—particularly, in this case, in the discourse of theoretical physicists.

Your idea is truly captivating. There must be some truth in it somewhere.
—Albert Einstein, letter to Theodor Kaluza, December 9, 1921

CHAPTER TWO

THEORIES FROM ANOTHER DIMENSION

Albert Einstein's Ground-Shattering Revolution and Theodor Kaluza's Radical Response

Mathematics is a curious kingdom, situated somewhere between the physically real and the utterly fantastic. Some formulations map beautifully onto nature, such as Maxwell's sleek unification of electricity and magnetism into electromagnetism by means of a basic set of equations (simplified even further by English mathematical physicist Oliver Heaviside). It resulted in the stunningly spot-on (and eminently physical) prediction of the speed of light. Other mathematical inventions, such as the one-sided surface of a Klein bottle—named for Felix Klein, a professor at the University of Göttingen who proposed the idea—are physically impossible in ordinary space. In higher dimensions, though, Klein bottles thrive—interiors linking to exteriors by means of a hyperspace twist.

Throughout the late nineteenth and early twentieth centuries, Göttingen's math department was a workshop for higher-dimensional geometry in general—crafting assorted constructs, many with initially unknown application to the physical world. In that refined milieu, young thinker Theodor Kaluza would spend a pivotal year of his life, starting in fall 1908. There, he'd develop a passion for the fifth dimension and attempt—about a decade later—to apply it to modern physics.

Ambitiously, Kaluza sought to unify the then-known forces of nature—gravitation and electromagnetism—by extending Einstein's general theory of relativity by an extra dimension in a manner that cleverly reproduces Maxwell's equations. Maxwell's unification of electricity and magnetism was so predictive and alluring, and Einstein's theory so mathematically elegant—based on natural geometric relationships—he wished to represent the two interactions on common ground. Space-time, though, didn't seem to offer enough room to craft his model, leading him to add one more dimension—in similar manner to attaching a new wing to a cramped office building to make it more functional. He was stunned when his five-dimensional theory seemed, at first, to forge unprecedented unity. Sharing his results with Einstein, Kaluza would spark more than a century of efforts to bring the natural forces together—including the weak and strong nuclear forces—under a higher-dimensional umbrella.

The connection between extra-dimensional and multiverse models is complicated. Early attempts at five-dimensional unification, by Kaluza and others, did not posit multiple realms. Rather, they hypothesized unseen dynamics within our own, single universe. Only in the late twentieth and early twenty-first century would ten- or eleven-dimensional string and M-theories offer string landscape and brane-world models that, in some cases, include the possibility of other universes.

FOUR-DIMENSIONAL THINKING

Given the late nineteenth-century association of higher dimensions and ether with spirituality and mysticism, perhaps it is not surprising that early twentieth-century proponents of realms beyond space were cautious. In crafting the special theory of relativity, which brilliantly resolved the discrepancy between the predictions of Newtonian and Maxwellian physics regarding the constancy of speed of light for all observers, Einstein set aside the need for ether altogether. Instead, he abolished absolute space and time, replacing them with relative constructs. From the perspective of a terrestrial observer, a spaceship moving close to the speed of light would contract along the direction of its motion. Clocks within it, from an earthbound viewpoint, would move more slowly. In tandem, such length contraction and time dilation would ensure the constancy of light speed.

Initially, Einstein also didn't make reference to higher dimensions—sticking with the conventional three dimensions of space and one dimension of time. The overall theme of his choices was to stick to the tangible, in accordance with the philosophy of Mach, whose work he greatly admired.

Hermann Minkowski, who had been Einstein's mathematics professor at ETH (the German-language initials for the Swiss Federal Institute of Technology), the university from which he had received his physics degree, had likely sensed his lack of enthusiasm for abstract math. Einstein had missed many of Minkowski's classes. Imagine Minkowski's amazement when his rogue student, who, upon graduation, had difficulty finding an academic position and landed a job at a Swiss patent office, developed a revolutionary theory of nature. By then, Minkowski had moved to Göttingen—the hub of higher-dimensional discourse. There, he discovered that he could smartly express Einsteinian special relativity—how spatial distances and temporal

Image 4. German-born physicist Albert Einstein, pioneer of quantum physics and developer of special and general relativity. Credit: AIP Emilio Segrè Visual Archives, W. F. Meggers Gallery of Nobel Laureates Collection, with thanks to the Albert Einstein Archives at Hebrew University.

intervals depend on the speeds of observers—by combining space and time into a unified, four-dimensional entity called space-time.

In September 1908, at the eightieth assembly of the Society of German Natural Scientists and Physicians, Minkowski announced his finding in a rather dramatic fashion:

> The views of space and time which I wish to lay before you have sprung from the soil of experimental physics, and therein lies their strength. They are radical. Henceforth space by itself, and time by itself, are doomed to fade away into mere shadows, and only a kind of union of the two will preserve an independent reality.[1]

Image 5. German mathematician Hermann Minkowski, proposer of four-dimensional space-time as a natural way of expressing special relativity. Credit: H. A. Lorentz, A. Einstein, H. Minkowski, *Das Relativitätsprinzip* (1915), courtesy of AIP Emilio Segrè Visual Archives, Born Collection.

For several years, Einstein was unimpressed by Minkowski's synthesis. By rewriting relativity in abstruse mathematical language, Einstein felt that his former lecturer was making straightforward physics unnecessarily esoteric—like rewriting the breakfast menu at a roadside diner in the form of a Shakespearean sonnet. Minkowski's reformulation seemed as rarefied and pointless to him, no doubt, as the higher math classes at ETH. He derided Minkowski's conjecture as "superfluous learnedness," and wrote with his collaborator Jakob Laub that it "makes great demands on the reader in its mathematical aspects."[2]

THEORIES FROM ANOTHER DIMENSION

Moreover, like many scientists who grew up in the late nineteenth century, Einstein considered the concept of higher dimensions an irrelevant abstraction at best, and a pretext for pseudoscience at worst. The term "dimension" he felt, made people shudder with its connotations of séances, spiritualism, the afterlife, and occult discussions in general. Einstein's sensitivity to those supernatural associations were such that, later in life, when a reporter asked him to define the fourth dimension, he misunderstood the journalist's intention, laughed, and replied, "ask a spiritualist," before conceding that it could be "time."[3]

By around 1911, however, convinced by his colleagues of its superior way of encapsulating special relativity, Einstein began to embrace Minkowski's four-dimensional fusion. From that point on, he'd tell others that there was nothing mysterious or convoluted about the fourth dimension being associated with time. It is fortunate that Einstein changed his tune. Four-dimensional differential geometry (a branch of higher mathematics) proved essential to his 1915 proposal to replace Newtonian gravitation with the general theory of relativity. In it, matter and energy warp the fabric of space-time, steering the orbits of celestial bodies accordingly.

General relativity explains, for example, why Earth—trapped in the sun's gravitational well—travels in an elliptical orbit around it. It also leads to bold predictions, such as the bending of starlight by massive objects such as the sun. Expeditions launched during the 1919 solar eclipse to Principe, an island off the coast of west Africa, and Sobral, in Brazil, splendidly supported Einstein's conjecture as superior to forecasts based on Newtonian gravitation. The power of Einstein's mathematical machinery would also drive a revolution in cosmology, brilliantly modeling the dynamics of the universe. Relativity—both the special and general theories—brought Einstein global fame, bestowed him with lifelong international press coverage, and made him the subject of much correspondence from other scientists eager to extend his work.

A GLORIOUS ARIA: THEODOR KALUZA'S MOMENT OF REVELATION

In April 1919, while living and working in Berlin under several key appointments, including professor at the University of Berlin and member of the Prussian Academy of Sciences, Einstein was intrigued to receive a letter from Kaluza, who was working then as a *privatdozent* at the University of Königsberg (then part of East Prussia, now Kaliningrad, Russia). A *privatdozent* is an instructor earning his pay by selling lecture tickets—a significantly lower rank than professors. Nonetheless, Einstein didn't care about status. Kaluza presented a detailed mathematical scheme to add another dimension—albeit an undetectable one—to space-time in an expanded version of general relativity. Because Maxwell's equations naturally emerge, gravitation became wedded to electromagnetism in a single unified theory.

Despite his poorly paid academic position, Kaluza was brilliant and came from a distinguished background. Born in Wilhelmsthal, Upper Silesia (then a German-speaking region of Prussia, now part of Poland), in 1885, his family soon moved to Königsberg, where his father, Max Kaluza, was appointed to a university position to study the works of Chaucer and other English literature. Thanks, in part, to his father, Theodor learned English in his youth. A polyglot, he went on to learn more than a dozen other languages, from Hungarian to Arabic. He spoke seven languages fluently, and ten others with some ability.[4] Persistent and clever, he mastered swimming merely by reading a book on the subject and taking the plunge—succeeding on his very first try.[5]

Mathematics, however, was Kaluza's prize subject. After completing his habilitation (permission to teach) in that subject in 1909, he was appointed to the University of Königsberg. He was only twenty-four at the time, sporting a bushy, coal-black beard—looking more like a bohemian student than an instructor. By report, he wasn't a particularly good teacher. Often distracted, on at least one occasion he forgot about class and headed to the movies instead, where some of his students eventually found him.[6]

Given Kaluza's interest in higher mathematics, and his earlier exposure to special relativity—much talked about during his year at the University of Göttingen due to Minkowski and others—general relativity captivated him. Alas, shortly after the publication of Einstein's key paper introducing that subject, he was called to battle in World War I. By the time he returned in 1918, some of the main problems in that field, such as the gravitational field surrounding static, spherical stars (and black holes), had been solved. Nevertheless, he knew that general relativity was in some sense incomplete because it described only gravitation, but not the other natural force known at the time, electromagnetism.

By that time, Kaluza had been married for more than a decade to Anna Beyer, with whom he had a son, Theodor Jr., and a daughter, Dorothea. Theodor Jr. loved to hang out with his father in his study, and sometimes watch him while he was completing his calculations. On one memorable occasion, the boy, then ten years old, was startled to see his father act rather oddly. As he later recalled:

> [My father] sat completely still for several seconds, and then he whispered very sharply and banged the table, and he stood up but remained completely motionless for several seconds. Then he began to hum the last part of an Aria of Figaro.[7]

That strange behavior constituted the elder Kaluza's eureka moment. He realized that he'd found a way to enlarge the four-dimensional tensors (mathematical entities that transform in particular ways during coordinate-system changes) used in general relativity to describe gravity into five-dimensional tensors that included new components pertaining to electromagnetism. General relativity's four-dimensional tensors are represented by 4x4 arrays of numbers or variables, for a total of sixteen entries each. By expanding those into five-dimensional tensors, represented by 5x5 arrays with a total of twenty-five entries, Kaluza added nine more components. For reasons of symmetry, only five of the new terms were independent of each other. Nicely, those extra entries carry sufficient information to characterize electromagnetic interactions. The process he used was akin to adding an extra row and column to a spreadsheet with the goal of incorporating more data. Just as adding a fifth row and fifth column to a tax spreadsheet might help include the details for a new dependent and explain a new line of expenses, respectively, Kaluza found that the additional space in the tensors of general relativity offered enough room to accommodate an extra force. Amazingly, processing those tensors through Einstein's mathematical machinery effortlessly reproduced Maxwell's equations, theoretically allowing for the unification of gravitation and electromagnetism into a single five-dimensional theory.

Realizing that no one has ever directly observed a fifth dimension, Kaluza employed a mathematical trick to make sure his theory reflected that truth. He included an ad hoc mandate called the "cylinder condition" that insisted that nothing physically measurable might include components of the extra dimension. That way, its effects might never be seen directly. The reference to "cylinder" stems from the notion that the extra dimension becomes cyclical—something like helicopter blades whirling around so quickly that you could never pin down their exact location.

Boldly, after writing up his results in a paper, "On the Unity Problem of Physics," Kaluza mailed it to Einstein. His fervent hope was that it would be included in the *Proceedings of the Prussian Academy of Sciences*, a highly respected journal, for which Einstein's recommendation would be critical.

Initially, Einstein was impressed by Kaluza's hypothesis. As he wrote back:

> The idea that [unity of the forces] can be achieved by a five-dimensional cylinder-world has never occurred to me and would seem to be altogether new. I like your idea at first sight very much.[8]

Einstein's exuberance quickly faded, however, when he began to plow through Kaluza's calculations and found some difficulties. True, the paper reproduced Maxwell's equations by means of an extension of general relativity. But if one changed the coordinate system in certain ways, such as performing certain types of rotations, that property disappeared. Moreover, it seemed to him artificial that the "cylinder condition" singled out the fifth dimension, without a clear physical explanation of why it didn't apply to the others. Finally, the paper lacked any account of how a charged particle, such as an electron, would behave under the combined influence of gravitational and electromagnetic fields in the theory. Surely a unification model, Einstein thought at the time, should reproduce physical reality. (In his own unification models, Einstein would eventually abandon that aspiration, which proved to be a tall order.)

A week after his first response to Kaluza, Einstein wrote to him again:

> [On] the whole I have to admit that the arguments brought forward so far do not appear convincing enough. . . .[9]

Einstein suggested that he should try another journal, one that was more mathematical and accepted long papers. Generously, however, he offered to consider Kaluza's work again if it was shortened, clarified, and not placed in any other journal. That offer proved useful to Kaluza two years later, when, after Einstein realized that the paper had never been published, decided to recommend it for the *Proceedings*. It was finally published in December 1921. Few theorists took note at the time, however—perhaps because general relativity was so fresh and quantum mechanics yet to be developed. It would take the emergence of the latter, and comparison with the former, to generate greater interest in unification. The challenge of bringing gravitation into the quantum realm, with the two theories so different in their mechanisms, would ultimately motivate further thinking about Kaluza's bold idea for unity.

DOUGHNUT DIMENSIONS

Several years after Kaluza's publication, Oskar Klein (no relation to Felix) independently arrived at the idea of five-dimensional unity himself—placing it, however, within a quantum context. In those days, well before online journals and search engines, rediscoveries of ideas were commonplace. If the reintroduction of the notion was sufficiently distinct, sometimes both the originator and the second discoverer would get joint credit. Hence, the designation "Kaluza-Klein theory," rather than just "Kaluza theory." Incidentally, yet another five-dimensional unification model predated Kaluza's—a 1914 proposal by Finnish physicist Gunnar Nordström that was based on a gravitational model soon rendered obsolete by general relativity.

Klein was born in Mörby, near Stockholm, to a religious and scholarly Jewish family. His father was a venerated cleric, Slovakian-born Gottlieb Klein, the first chief rabbi of Stockholm. Oskar's mother, German-born Antonie "Toni" Klein (née Levy), had academic roots. He grew up fascinated by science, including biology and chemistry, and was pleased to perform research—starting at the age of sixteen, while still in high school, under the guidance of physical chemist Svante Arrhenius, the first Swedish Nobel laureate and director of the Nobel Institute. Arrhenius served as his mentor throughout his university career.

In 1918, the golden opportunity arose for Klein to travel to Copenhagen and work with Bohr. Bohr had started to build a quantum

Xanadu in that magical, northern city, beginning with rudimentary atomic models that resembled the solar system, with electrons as "planets," and culminating in full-fledged quantum mechanics with its probabilistic description of fundamental interactions. Klein was delighted to learn from the great thinker, and eventually serve as one of his leading research assistants. Throughout the late 1910s and early 1920s, he journeyed back and forth between the two Scandinavian capitals to perform research in the world-renowned centers led by Arrhenius and Bohr.

By the mid-to-late 1920s, a new haven for quantum physics had emerged at the University of Michigan in Ann Arbor, eventually attracting notables such as Samuel Goudsmit and George Uhlenbeck, developers of the concept of quantum spin: the notion that particles such as electrons have discrete amounts of intrinsic angular momentum that govern how they react to magnetic fields. There, Klein landed a teaching position, starting in fall 1923. While enlightening students about electromagnetism the following year, he began to consider the question of what would happen to electrons under the combined influence of electromagnetic and gravitational fields.

In 1924, French physicist Louis de Broglie published his brilliant theory of matter waves. He showed how to understand the behavior of electrons in atoms by using a standing wave analogy. Standing waves are the distinct patterns in an oscillating string (or similarly flexible cord of material) attached on both ends and set into motion. To fit exactly within its boundaries, such constricted strings must have an integer number of peaks: 1, 2, 3, or more. Similarly, an electron's waveform, as it occupies the circumference of a loop around the nucleus, is restricted to an integer number of peaks. That value, he found, well matched the previously developed "primary quantum number," which tagged each orbit of Bohr's "solar system" atomic model. German physicist Arnold Sommerfeld had enhanced Bohr's model in the mid-1910s by adding two more quantum numbers associated with angular momentum. Quantum spin, introduced in 1925, would complete the picture with a fourth quantum number. Those four quantum numbers delineated the quantum states of electrons in atoms. Representing electrons as standing waves keyed to the primary quantum number helped flesh out that idea.

De Broglie's matter wave hypothesis stimulated many young researchers around the world to think of electrons in a new light, as having both wave and particle properties, and using those twin aspects to justify their behavior. Klein became drawn to the question of justifying the discreteness (multiples of a finite, constant value) of electric charge, by relating it to a wave property of electrons. As de Broglie had shown, the linear momentum of an electron—a particle property—was inversely proportional to its wavelength—a wave property. Neither of those explained the discreteness of charge, however.

Recalling an offhand remark by Bohr that the fourth dimension might be insufficient to explain the particle world, Klein took a mental leap into the fifth dimension. He discovered a neat way to connect the charge of electrons with the wave properties of a five-dimensional ring in a unified theory of electromagnetism and gravitation. In doing so, he developed a unified theory similar to Kaluza's but with the key difference that the extra dimension, rather than being shielded from direct observation by arbitrary mathematical constraints, would be wrapped up instead into a tight circle, roughly the size of what is called the "Planck length."

The Planck length is one of the natural units introduced by German physicist Max Planck in 1899 in an effort to regularize definitions of physical properties. Approximately 10^{-35} meters, it is about as minuscule as you can get in theoretical physics—far, far smaller than anything directly measurable. It represents the scale at which, for a glob of charged matter, gravitational and electromagnetic forces would be comparable in strength. At more mundane scales, gravity is much, much weaker. It also represents the domain in which quantum effects would likely make their presence known in gravity, much as they do in electromagnetism at far larger scales. Because Planck-length-sized phenomena would be undetectable, Klein saw no issue with a fifth dimension of that scale.

Imagine an aerial view of a gardening supply yard with a set of straight, thin rubber tubes, some solid and some hollow, scattered over a patio. From high above, the tubes would seem one-dimensional, like sticks, and it would be hard to say anything about their interior. The dimension corresponding to the cross sections of the tubes would be very hard to observe. Yet the hollowness of some of the tubes enables them to have the practical function of being able to convey water, as garden hoses. Similarly, while in Klein's unified theory the fifth dimension is experimentally undetectable, it constitutes an integral part of the theory's mission.

The specific way Klein justified why electron charge comes in discrete multiples of a fixed value is to connect it with five-dimensional momentum. He generalized de Broglie's procedure, in which momentum is inversely proportional to wavelength, to argue that five-dimensional momentum is inversely proportional to a five-dimensional wavelength. Then, if electrons are standing waves in a circular fifth dimension, of Planck-length size, integer multiples of the wavelength must fit into the circumference. That forces the five-dimensional momentum, and hence the charge, to be integer multiples of a fixed value—justifying the quantization of electric charge.

Klein was very enthusiastic about the idea at first. He entranced his colleagues by talking about its potential to link together all of modern physics. As Uhlenbeck, who knew him well, recalled:

> I still remember one time after these discussions with Klein in which he had told about his five-dimensional relativity and how out of that quantum conditions would come. You see, from the periodicity condition in the fifth dimension you got the quantum conditions. And I was so excited. I told them [my friends],

"Very soon we . . . [will have] the world formalized. We will know everything! Everything will be known at that time." Well, it was a beautiful exaggeration.[10]

After Klein returned to Copenhagen for a 1926 visit, he was in for a shock. Physicist Wolfgang Pauli, who often took pleasure pointing out the flaws in others' theories, learned about his work and bluntly informed him about Kaluza's earlier idea and its similar reliance on an undetectable fifth dimension. Klein looked up Kaluza's paper for the first time, and felt like he was hit by a freight train. The notion that all his hard work merely duplicated another's idea that he could have looked up easily if he had been pointed to the right place resounded on repeat in his tired brain. Once he managed to collect himself, however, he realized that there were substantive differences between the two papers: not just his mechanism for wrapping up the fifth dimension in a tiny circle, but also the exactness of some of his calculations compared to some places where Kaluza had approximated. Satisfied that his work was sufficiently original, Klein decided to publish.

What physicists today call Kaluza-Klein theory has ended up preserving Klein's concept of curled-up extra dimensions. In the interim, as the number of known fundamental forces has doubled from two to four—supplementing gravitation and electromagnetism with strong and weak nuclear interactions—contemporary theories that include Kaluza-Klein mechanisms typically incorporate six extra, tiny curled-up dimensions in addition to ordinary space-time, and, in some cases, also include a large (not curled up) extra dimension that is inaccessible for other reasons, for a total of ten or eleven dimensions. These more modern theories include supergravity, superstring theory, and M-theory.

The generalization of a ring into a higher dimension is a doughnut-shape, or torus. A three-dimensional torus is a composite of two perpendicular circles: an outer ring marking the extent of the doughnut if it is placed on a plate and an inner ring corresponding to the circumference of its cross section; that is, how it would be sliced. Tori can be defined in any number of dimensions by adding more rings perpendicular to the others.

The process by which some, but not all, of the dimensions in a manifold become curled up is called "compactification." There are many such ways that might occur, with minute, multidimensional doughnuts constituting but one of many possibilities. More commonly today, compactification leads to what are called "Calabi-Yau manifolds," higher-dimensional spaces with special geometric properties, named for mathematicians Eugenio Calabi, who posited their existence, and Shing-Tung Yau, who confirmed Calabi's hypothesis and defined their features. Twisted up in the higher dimensions into various contortions, they are more like pretzels or balls of twine than doughnuts. Their mathematical features lend themselves well to superstring models that include the various symmetries of quantum field theories describing the strong and electroweak (electromagnetism combined with weak) interactions. Growing under the challenge of a far grander unification than once anticipated, Kaluza-Klein theory has certainly blossomed well beyond its original seedlings.

DOES GOD ROLL DICE OR PLAY STEALTHY POKER INSTEAD?

Einstein maintained a passionate interest in unified field theories—including several different varieties of higher-dimensional extensions of general relativity in the vein of Kaluza and Klein—throughout the final three decades of his life. In his main Kaluza-Klein efforts in the late 1930s and early 1940s, his stalwart assistants were researchers Peter Bergmann and Valentine "Valya" Bargmann. Even on his deathbed, in April 1955, Einstein asked for a pencil and notepad to attempt some calculations with the goal of steps toward unification.

Motivating Einstein was his feeling, starting in the mid-1920s, that quantum physics had veered along the wrong path, partly because of its incorporation of probabilistic element. That new direction emerged mainly due to the pivotal contributions of German physicists Max Born and Werner Heisenberg, who made likelihoods, rather than certainties, key aspects. To emphasize its novel mechanisms including chance transitions between various electron states, Born coined the term "quantum mechanics" as a replacement for the deterministic laws of Newtonian physics. In accordance with his belief in the philosophy of Dutch thinker Baruch Spinoza, who had equated God with an optimal state of nature, Einstein argued that a complete theory of reality must be seamless in its connections between past, present, and future. Chance results could not be fundamental, he thought. Rather, their inclusion called for a deeper, more comprehensive theory than quantum mechanics. As he wrote to Born in December 1926:

> Quantum mechanics is very worthy of respect. But an inner voice tells me this is not the genuine article after all. The theory delivers much but it hardly brings us closer to the Old One's secret. In any event, I am convinced that He is not playing dice.[11]

Einstein's repeated ventures into attempted higher-dimension unification models of various types—some more abstract, others more physical—showed that he was willing to let God play poker instead of dice, keeping His cards hidden at times. Anticipating the tradeoff made by many modern crafters of multiverse models, he was more than willing to introduce unobservable components into his theories, such as an undetectable fifth dimension, with the aim of keeping nature's garment stitched together at any cost.

Often, Einstein would be met by snide critiques from Pauli, one of the few major physicists who kept up with his late-career unification efforts. The flaws Pauli unearthed in Einstein's conceptions would generally prove lethal. Undaunted, he would simply move on and attempt another method. Along with other ways of modifying general relativity, adding a fifth dimension served as one of the many tools in his toolbox.

In a sense, Einstein was harking back to Plato's way of explaining apparent imperfection in the world. Plato postulated a realm of what he called Forms as the idealization of what we see on Earth. Each Form is the perfect version of some Earthly experience, such as ideal love and immortal life. Mundane doings are meager shadows of the goings-on in the flawless, otherworldly reality.

As an analogy, Plato imagined a situation in which long-term prisoners are shackled inside a cave, near enough to its entrance that they might observe shadows on the wall of passersby, but still unable to see outside. As merchants, warriors, and others walk past the cavern's opening, the prisoners see only the dusky two-dimensional projections. If they've been confined so long that they have forgotten about the outside world, they might well mistake the shadows for the real thing. Accordingly, Plato argued, we might mistake our flawed world for reality, when it is only a projection of the realm of Forms. Similarly, Kaluza, Klein, and Einstein (during his forays into higher-dimension unification models) each imagined that the observed, four-dimensional physical domain is merely a shadow of a true five-dimensional reality.

The idea of unseen, extra dimensions means that there are neighboring spaces only a minuscule distance away from us at all times. Even if they are wholly inaccessible, that is truly mind-blowing. Philosophically, one might wonder, is that too dear a price to pay to avoid uncertainty in nature? Einstein, at various points in his career, didn't seem to think so, despite being far more of a realist in his youth.

In a game of poker, the card deck and the number of its possible orderings are both finite. Once it is shuffled and the hands dealt, although the cards are unseen by all but the player holding them, their specific distribution offers a deterministic element operating behind the scenes. The arrangements of cards with certain ranks and suits—along with the skill of the players—govern the game's progress and outcome. Therefore, although the individual players, not knowing the ranks and suits of the cards that are facing down, would not have enough information right after the hands are dealt to predict what will happen, someone who happened to know the exact order of the deck right after it was shuffled (by cheating using marked cards, for instance) would be in an excellent position to predict the outcome. Dice-rolling using fair dice, on the other hand, would not allow anyone to make such predictions.

Along such lines of reasoning, Einstein preferred models with hidden dynamics, such as five-dimensional extensions of general relativity, to those with purely random mechanisms, such as the orthodox interpretation of quantum mechanics. In the former case, at least prognostications would be theoretically possible. Of course, a completely open, predictable model would be best of all, but he was unsure if that was possible.

EINSTEIN'S MOUSE

Einstein's inquiries about the quandaries of quantum mechanics helped inspire another type of multiverse model that is perhaps even stranger than hidden dimensions: the "Many Worlds Interpretation," as it later would become known, developed in the mid-1950s by Hugh Everett, a young graduate student at Princeton. To keep nature fully deterministic and avoid quantum "dice-rolling," rather than extensions into a higher dimension, it posits a frequent branching of reality into parallel strands.

Everett likely was in the audience of the last lecture delivered by Einstein, on April 14, 1954, at Princeton, and heard the founder of relativity point out what he saw as a fatal flaw in the orthodox interpretation of quantum measurement theory. By then, de Broglie's matter waves depicting sinuous electrons had long been replaced by more abstract "wave functions" representing probability distributions of particle properties for each quantum state. Austrian physicist Erwin Schrödinger's wave equation, proposed in 1926, showed how those wave functions evolve continuously over time up until the moment a researcher decides to measure a value. According to orthodox views, as formulated in particular by mathematician John von Neumann in the late 1920s, at the instant of measurement, "wave function collapse," the process by which a wave function representing many options narrows down into one signifying a single definitive result, transpires. Einstein didn't see why, as Bohr, von Neumann, and others had emphasized, a human observer should enter into the picture. Nature should operate seamlessly on its own.

For Einstein that meant finding a deterministic, unified field theory that transcended and hence explained probabilistic quantum mechanics. Everett would decide to take a different route, keeping the deterministic elements of standard quantum mechanics, such as Schrödinger's wave equation, while setting aside the need for observers to precipitate quantum measurement processes. The evolution of the wave function would simply continue, and the observers would constitute blends—as strange as that sounds—of a range of quantum outcomes.

Einstein's final talk was organized by Princeton physicist John Wheeler, who played a major role in the renaissance of general relativity as an active discipline in the 1950s and 1960s. In a chicken and egg situation, Wheeler learned by teaching and taught based on what he picked up. It is hard to say which came first: him starting Princeton's first course in relativity or his interest in pursuing the subject as a research topic. One fed into the other. At any rate, his desire to plunge into general relativity, both in teaching and research, began after he returned to Princeton in the early 1950s after contributing to military research at Los Alamos and elsewhere working on the hydrogen bomb. He started Princeton's first full-year course in relativity, and began to explore the mysteries of gravitation in his scholarship.

Because Wheeler knew Einstein well—they resided a few blocks away from each other—he invited the founder of relativity to deliver a guest lecture to the small group of students taking his course on the subject. Everett likely learned about the talk from his close friend Charles Misner, a PhD student of Wheeler. Due to Einstein's fame, Wheeler kept the talk under wraps lest it attract a public stampede. Einstein could speak about virtually anything, and it would make headlines. As it turned out, at the time of the seminar, the genius physicist was in his final stage of life. He would die little more than one year later of a heart aneurysm.

In his lecture, after discussing aspects of general relativity, Einstein turned to the topic of quantum measurement. He expressed his incredulity that human observation precipitated wave function collapse. If so, why just people? Why not animals or beings? Shouldn't fundamental natural processes be automatic, not species-dependent?

As Wheeler recalled, Einstein asked the group rhetorically: "When a person such as a mouse observes the universe, does that change the state of the universe?"[12]

Shortly thereafter, Everett would craft his own solution to the quantum measurement problem: a universal wave function encompassing the entire cosmos that ties the fates of all things—including researchers and any phenomena they record—into a single, ever-flowing entity. It wouldn't matter if people, mice, or nothing at all were observing an experiment. If there were two options for the outcome of a measurement, there would be two nearly identical copies of any witnesses—one version that perceived the first possibility and another that saw the second.

In making that bold proposal, Everett, while still a graduate student, positioned himself in the opposing camp to quantum luminaries such as Bohr and von Neumann. It was an uneven battle for sure. Nonetheless, Everett's radical perspective has continued to pick up adherents in the intervening decades—those looking for alternatives to quantum orthodoxy. It has also launched interest in multiverse ideas in general. After all, if reality might have unseen dimensions, postulating unobservable parallel universes does not seem any more of a stretch. Moreover, while both scenarios sound weird, they might not be any more bizarre than human scientists playing essential roles in quantum dynamics.

One can imagine an intelligent amoeba with a good memory. As time progresses the amoeba is constantly splitting, each time the resulting amoebas having the same memories as the parent. Our amoeba hence does not have a life line, but a life tree. . . . It becomes simply a matter of terminology as to whether they should be thought of as the same amoeba or not, or whether the phrase "the amoeba" should be reserved for the whole ensemble.

We can get a closer analogy if we were to take one of these intelligent amoebas, erase his past memories, and render him unconscious while he underwent fission, placing the two resulting amoebas in separate tanks, and repeating this process for all succeeding generations, so that none of the amoebas would be aware of their splitting. After a while we would have a large number of individuals, sharing some memories with one another, differing in others, each of which is completely unaware of his "other selves" and under the impression that he is a unique individual.

—Hugh Everett, from an early draft of his 1957 doctoral dissertation, "On the Foundations of Quantum Mechanics"

CHAPTER THREE

SHOWDOWN IN HILBERT'S HOTEL

The Competing Quantum Visions of Niels Bohr, Hugh Everett, and Others

Battling the scientific establishment takes courage, especially if the defender is almost universally revered and the contender is a graduate student in his mid-twenties. Niels Bohr was a champion to the world of physics and a hero to the nation of Denmark, tantamount to scientific royalty. Though he spoke in a quiet murmur, he was a figure whose statements would be cherished and analyzed. When British monarch Queen Elizabeth II paid a state visit to that country, in spring 1957, meeting with Bohr was a key stop on her itinerary.

The young thinker who would challenge Bohr's orthodoxy, Hugh Everett III, was taciturn, headstrong, and distrustful of authority. Though cynical, he was bright enough that his fascination with the world, and knack for reasoning, carried him through his darker moments and gave him purpose in life. On the flip side, his brilliance sometimes led him to feel singular and superior, filling him with disdain for the foibles of others.

Perhaps it was inevitable that someone with Everett's temperament would be the one to point out that quantum measurement theory, as it stood, was "an emperor with no clothes"—to reference a story by another famous Dane, Hans Christian Andersen. Some critics would later consider Everett's proposed multiverse solution to quantum dilemmas similarly a fairy tale, but that's another story. Courage means the freedom to speak out, rightly or wrongly, as the case may be. At the very least, Everett's bold challenge opened the floodgates of discussion, ultimately rendering Bohr's traditional view but one of many alternatives under consideration.

For his undergraduate studies, Everett had attended the Catholic University of America, where his penchant for intellectual combat soon became apparent. Anticipating his radical new interpretation of quantum physics, he enrolled in a philosophy of science course, for which he received an A. Irreligious himself, he relished debating the existence of God with the clerical professors teaching the required courses in theology. He learned about classic ontological (philosophy of existence) proofs of God's necessity and—much to his teachers' discomfort—constructed his own rebuttals, one applying similar logic to argue for the existence of the mythological horse Pegasus.[1]

In delving into ontological arguments, Everett may have come across the famous "possible worlds" line of reasoning advanced by eighteenth-century philosopher Gottfried Leibniz—a rival to Newton in the controversy as to who invented calculus. In his 1710 treatise *Theodicy*, Leibniz argued that a perfect God would envision every feasible variation of the universe and choose from among them "the best of all possible worlds." Those alternative realities would exist only in the mind of God, who would judge them based on their fitness, and bring the finest into being. Without such selection, there would be a myriad of parallel realities—a multiverse, in modern parlance—each operating on its own terms for better or worse, none being more special than the other. In other words, without the benevolent filter of the divine the cosmos would be pure pandemonium. Regardless of whether or not Leibniz may conceivably have been an influence, it is interesting to note that Everett's future work would remove subjective observation, and arguably lead to such a chaos of unrestricted multiple worlds.

A NEXUS OF GENIUS

In 1953, Everett began a graduate program in physics at Princeton. The prospect of crossing paths with seminal figures in science undoubtedly induced many in those days to flock to that leafy, intellectual town. For him, it would be a chance to learn from such icons, but also be unafraid to point out any inconsistencies in their thinking.

It had been two decades since the Institute for Advanced Study (IAS) was founded in the town of Princeton as an independent academic establishment close to the university. From the start, it housed two of the greatest geniuses in the world. One was Albert Einstein, who took up a permanent position at the IAS beginning in October 1933 partly due to threats on his life by the monstrous Nazi regime that had taken power in Germany. The other was Hungarian émigré scientist John von Neumann. Many more extraordinary minds would join them—including the brilliant Austrian logician Kurt Gödel—but Einstein and von Neumann arguably made the most impact on the widest range of scientific inquiry.

About six years earlier, starting around 1927, von Neumann had begun to lay the mathematical foundation for what became known as the "Copenhagen interpretation," or orthodox interpretation, of quantum mechanics. One of its hallmarks was an infinite-dimensional abstract quantum space, which he dubbed "Hilbert space" after the mathematician

Image 6. Hungarian-American mathematician John von Neumann, who codified the orthodox interpretation of quantum mechanics involving two separate processes: continuous evolution and measurement-induced collapse. Credit: AIP Emilio Segrè Visual Archives.

David Hilbert, who had made many strides in fundamental algebra theory. Unlike the (four or five) dimensions explored by Einstein, Minkowski, Kaluza, and Klein, in relativity and its generalizations, the abstract avenues of Hilbert space did not directly contain matter or energy, rather they included innumerable quantum states indirectly representing every possible outcome of any given measurements.

Whimsically, as relayed by physicist and science writer George Gamow, Hilbert humanized the concept of infinity through the analogy of a hotel with unlimited rooms. Such a lodging would always have vacancies, even if all the rooms are full. If a wayfarer approached the front desk with a need for accommodation on an evening when the hotel was packed, the reception clerk could simply reassign the occupant of Room 1 to Room 2, that of Room 2 to Room 3, and so forth. Thus, Room 1 would be freed up for the last-minute guest. Even if an infinite number of guests needed rooms, the clerk could reassign all the existing occupants to even-numbered rooms (Room 1 relocates to 2; 2 to 4; 3 to 6; etc.), freeing up the odd-numbered rooms for the newcomers. In short, Hilbert's hotel would never have to turn anyone down.

Similarly, Hilbert space, with its potential for quantum states encompassing components spanning an infinite number of dimensions, is as flexible as can be. Von Neumann found it the perfect venue to house his detailed mathematical description of quantum reality.

ANTHROPOGENIC COLLAPSE: HUMAN OBSERVERS AND QUANTUM PROCESSES

Humans are blamed for many types of natural devastation. Yet, according to some scientific interpretations, our propensity to cause collapse may play an essential role in quantum mechanics—a type of disruption we needn't feel guilty about.

In his groundbreaking work in the late 1920s, von Neumann divided quantum processes into two distinct mechanisms: one deterministic and the other—to Einstein's chagrin—governed by chance and human measurement. The first process, the evolution of a quantum state over time following a wave equation developed by Austrian physicist Erwin Schrödinger, represented reality as a continuous strand from past to present to future.

Schrödinger's equation, published in 1926, was designed to explain how de Broglie matter waves assume particular configurations and evolve over time—showing how electrons, for instance, settle into their energy levels in atoms—but was later repurposed by Max Born and others to explain how probability waves evolve over time. These probability waves, called "wave functions," harbor information about the chances of particles having certain properties. Therefore, Schrödinger's equation tracks the progression of the probabilistic wave function, not the precise behavior of particles.

Throughout that evolution, because quantum states are represented by probability waves, physical quantities such as position and momentum would not, in general, be well-defined. For example, the quantum state representing an electron might reflect a range of possible positions involving a spread of likelihoods. Physicists would plot that as a wave function with a haze of values. That distribution might change over time—for instance if the electron is influenced by a changing voltage—but wouldn't naturally settle into a single amount. The same for momentum and

Image 7. Austrian Nobel laureate physicist Erwin Schrödinger, developer of the Schrödinger wave equation, the Schrödinger cat paradox, and many other scientific and philosophical contributions. Credit: AIP Emilio Segrè Visual Archives, Physics Today Collection.

other observables: they'd naturally continue to evolve as blurry mixtures, rather than resolving into sharp, pinned-down quantities.

To establish specific values of position, momentum, and other physical properties, that's when the second process, called "wave function collapse," kicks in. Unlike the first process, which is generally gradual (steadily turn a voltage dial and the wave function slowly morphs) quantum collapse is effectively instantaneous. It happens as suddenly and transformatively as the demolition of a grand, old building into an amorphous pile of rubble.

The notion of collapse depends upon the idea that any quantum state can be written as a superposition (a weighted combination) of component states in Hilbert space. Those building-block states, corresponding to unambiguous values of particular physical features, are known as "eigenstates." Depending on which physical property is under consideration, a quantum state can be expressed as a superposition of position eigenstates, momentum eigenstates, or eigenstates associated with other observables. The complete array of eigenstates associated with a particular property is called a "basis": a position basis, momentum basis, or otherwise. The choice of basis is akin to constructing a wood cabin with either all horizontal beams or all vertical beams, but not both types at once. Similarly, once a quantum state is expressed in a position basis, it constitutes a weighted combination of only position eigenstates. Alternatively, if framed in a momentum basis, it is a different weighting of only momentum eigenstates.

Now suppose a researcher implements an experiment designed to pin down the value of a certain observable (measured property), such as the position of an electron. According to von Neumann, as soon as its position is measured, the electron's mixed quantum state collapses randomly down into one of its position eigenstates, corresponding to an exact location. Although the collapse would be random, the chances of any particular outcome (a certain eigenstate and its position) are governed by the weighting of the original superposition.

While von Neumann's first process, quantum evolution according to the Schrödinger equation, is eminently reversible, his second process, quantum collapse, is irreversible in time, much like the fall of a house of cards. Videotape such a collapse, play it backward, and the reversed picture would look very unnatural. Nature obeys the Second Law of Thermodynamics (the principle that entropy never naturally decreases on its own), so such time-irreversibility seemed essential. Including human agents in the picture was odd for a physical process, but it well matched what was happening in the laboratory.

At the time von Neumann advanced his mathematical rendering of the Copenhagen interpretation, its principal advocate was the unchallenged spiritual leader of the field of quantum physics itself, heading one of the world's most prominent theoretical centers in that Scandinavian city. Bohr, director of the Institute for Theoretical Physics, had seen his star continue to rise. Debates with Einstein about the meaning of quantum physics cemented his role further as its greatest booster and sage.

Bohr called his quantum philosophy "complementarity," signifying a union of opposites akin to the ancient Chinese cosmological concept of yin-yang. Just as its symbol—which Bohr eventually included on his coat of arms—depicts a dichotomy of darkness and light, Bohr focused on the duality between the wavelike and particle-like properties of quantum experiments. Quantum physics, he emphasized, is a kind of black box with the type of results dependent

Image 8. Danish physicist Niels Bohr, recipient of the 1922 Nobel Prize in Physics, who made pivotal contributions to the development of quantum physics. Credit: Photograph by A. B. Lagrelius and Westphal, courtesy of AIP Emilio Segrè Visual Archives, W. F. Meggers Gallery of Nobel Laureates Collection.

on the decisions of the individual—a conscious observer—performing the measurement. For example, if an experimenter decides to record the position of an electron, the result is a specific position value, but if she chooses momentum instead, the position would remain vague.

Bohr's complementarity rivaled, and claimed to encompass, another shorthand for quantum weirdness: the uncertainty principle proposed by German physicist Werner Heisenberg in 1927. Heisenberg noted that when the quantum state of an elementary particle corresponds to a position eigenstate—that is, its position is known precisely—and one rewrites that quantum state in terms of a superposition of momentum eigenstates instead, that state is broadly smeared out over a vast distribution of momentum values. That is, if position is well known, momentum is very hazy. On the other hand, if momentum is pinned down, position becomes vague. Heisenberg codified that situation by means of an inverse relationship between the uncertainty in position and the uncertainty in momentum; the greater the former, the lesser the latter, and the converse. Heisenberg also identified energy and time as similarly paired quantities in terms of their uncertainties.

Unlike complementarity, which Bohr framed using intuitive arguments rather than math, Heisenberg's uncertainty principle emerged by applying mathematics to the properties of quantum states in Hilbert space. Nevertheless, aspiring to maintain harmony in the world of quantum physicists, Bohr had a tacit agreement with Heisenberg that complementarity and the uncertainty principle were essentially different ways of saying the same thing.

One seemingly paradoxical aspect of Bohr's quantum vision is that although humans, as observers, guide atomic and subatomic processes, they are made of atoms themselves. Such atomic components should therefore be subject to quantum rules too. But who would be the conscious measurers of those? To avoid such a conundrum, Bohr treated observers and the observed very differently, drawing a demarcation line between the two. In his view, the human measurer would effectively reside in a classical world, acting as a sentient outsider probing the hidden mechanisms of the atomic domain.

In general, von Neumann's notion of wave function collapse triggered by observation meshed well with Bohr's musings. It similarly embraced a cleave between the weird quantum rules of measured subatomic processes and the familiar laws human observers experience. After a hard day in the lab, researchers can relax, enjoy their dinners using Newtonian methods to cut their food, lift their beverages, and wipe their faces with napkins, and forget about quantum strangeness for a while. Von Neumann nicely lent greater mathematical clarity to Bohr's conjectures—creating, in tandem, the Copenhagen interpretation that was the standard for many years (and, even now, the most common explanation in textbooks).

Image 9. German Nobel laureate physicist Werner Heisenberg, developer of matrix mechanics, incorporated into quantum mechanics. Credit: Max-Planck-Institut, courtesy of AIP Emilio Segrè Visual Archives.

The two had very different lecturing styles. Bohr spoke softly, slowly, and enigmatically, often mumbling his words. Physicist Kenneth Ford recalled attending a lecture by Bohr, barely understanding what he said, and watching him become increasingly tangled up in his microphone cord as he repeatedly alternated between facing the blackboard and addressing the audience.[2] Von Neumann, in contrast, tore through his lectures at top speed—crystal clear but lightning fast. He and some of the other émigré Hungarian scientists were so otherworldly in their brilliance and energy level that they acquired the nickname, the "Martians." He made major contributions to so many important disciplines in addition to quantum physics, including computer science and game theory (a branch of economics connected with psychology), each unfurled with uncanny logic and clarity.

As mathematician Albert Tucker, another Princeton pioneer in game theory, noted about his speaking style:

> Von Neumann was so terribly quick in lecturing that people had to slow him up by asking questions. . . . [He] had a way of taking an idea that he had and explaining it very quickly and very clearly.

Unfortunately, von Neumann drove like he spoke. No trees were safe from his speeding and swerving. He raced through the world—figuratively from one problem to another, and literally in his smashed-up vehicles.

Bohr, in contrast, was no daredevil, especially as he got older. Rather, he became increasingly cautious, particularly in his physics. If he were to run into problems, it would be through absentmindedness, rather than recklessness. For instance, one time when speaking on a podium, he wanted to lower the lights. He pressed the wrong button, and the podium sank beneath the stage. Nevertheless, he kept speaking.

The combination of wariness to change and sometimes being out of touch with his environment made Bohr an ardent defender of the quantum gospel but—since his famous skirmishes with Einstein in the late 1920s and early 1930s—not a particularly good debater. When confronted about possible paradoxes and contradictions, he would often simply go silent.

The David to the Goliath of quantum orthodoxy was an unlikely figure. Young Everett decided to challenge the establishment with his radical PhD thesis at Princeton, questioning the necessity of von Neumann's second process, and brushing aside Bohr's artificial division between the quantum world of particles and the classical world of human observers. To characterize Everett's approach (echoing the words of cartoonist Walt Kelly): we have met the quantum, and it is us.

A VERY STUBBORN BOY

On November 11, 1930, in Washington, DC, a city known for its Neoclassical architecture, Hugh Everett III's classical world line in space-time began—in other words, in this timeline at least, he was born. His mother, Katherine Kennedy, was a talented poet who sadly suffered from mental illness. His father, Colonel Hugh Everett Jr., was an army officer and engineer, who had a horrible, sadistic streak. To teach a child to swim, for instance, he advocated throwing them—kicking and screaming—right into a lake (and unlike Kaluza's case, without having read an instruction manual). Starting at a young age, the Colonel ridiculed Hugh for being overweight, calling him "Pudge" instead of his given name. Hugh loathed that nickname, but

Image 10. Quantum physicists and others in conversation at Princeton Unversity. From left to right: Charles Misner, Hale Trotter, Niels Bohr, Hugh Everett III, proposer of the theory that became known as the Many Worlds Interpretation of quantum mechanics, and David Harrison. Credit: Photograph by Alan Richards, courtesy of AIP Emilio Segrè Visual Archives.

his dad didn't seem to care. Against his wishes, his father also insisted that he attend military school. His parents argued frequently, and divorced when he was only nine years old. Needless to say, Hugh felt bad for his mother, despised his father, and had a miserable childhood in general.

To her credit, Katherine encouraged Hugh to be creative. Undoubtedly because of her influence, he wrote some poetry himself. He also, at the age of twelve, penned a letter to Einstein. His letter is lost, but the gist of it might be inferred by the genius physicist's response. Apparently, he cheekily posed the age-old puzzle of what happens when an irresistible force meets an immovable body, thinking he might stump the great man. Without missing a beat, Einstein replied:

> Dear Hugh:
> There is no such thing like an irresistible force and immovable body. But there seems to be a very stubborn boy who has forced his way victoriously through strange difficulties created by himself for this purpose.
>
> Sincerely yours,
> A. Einstein

Suffering from the emotional scars of his formative years, Everett became headstrong and independent. In personal interactions, he was notably aloof—save, perhaps, when he imbibed. The family's choice for him to attend Catholic University as an undergraduate, despite his irreverence toward organized religion, stemmed, at least in part, from its proximity—being similarly in Washington. There, his mathematical gifts helped him sail through a demanding chemical engineering curriculum, including an impressive range of science and math courses.

By the time Everett arrived at Princeton in 1953, both Einstein and von Neumann were in their final years, as it turned out—the former with heart disease, and the latter soon to be diagnosed with terminal cancer. Also, neither was inclined to take on graduate students—which their membership in the IAS allowed them to avoid.

Everett gravitated toward Tucker, at first, as a mentor, hoping to learn more about game theory and offer a contribution to the field. While that discipline began with von Neumann's groundbreaking 1928 paper "Zur Theorie der Gesellschaftsspiele (On the Theory of Parlor Games)," describing the mathematics of zero-sum games, it truly took off with the atomic era and the start of the Cold War. In that dangerous age, in which devastating nuclear war could have been triggered by a mere miscalculation, research centers at universities and private think tanks such as the RAND Corporation explored various conflict scenarios, with rewards and penalties contingent on certain types of behavior, to gauge likely outcomes.

Tucker combined a competitive situation posed by Merrill Flood and Melvin Dresher of RAND with a narrative involving criminals in a gang being offered incentives to snitch on each other. Their combined sentences would be highest if both serve as informants, bolstering the government's case, and lowest if each remains silent. However, if one snitches and the other doesn't, the betrayer would benefit by being set free, while the other would be jailed for a long time. Tucker dubbed the model, the "Prisoner's Dilemma."

Along with conducting his own research, Tucker enriched the atmosphere at Princeton by running a colloquium series in which notables would present. Everett could well have remained in that ambience if it weren't for several factors that would steer him toward trying to reform quantum measurement theory by removing the role of the observer. One reason he switched was his interest in von Neumann's quantum work encountered via the textbook used in a course on that subject. Another was the influence of a circle of friends he met at Princeton, who opened him up to novel scientific avenues of exploration.

A HAZE OF HISTORIES

Princeton's Graduate College is a castle-like enclave unto itself, separated from the main campus by enough physical space to create a psychological distance. With its Gothic buildings with slate roofs, comical gargoyles, sturdy stone walls and towers, spacious courtyards surrounded by dormitories, and robed ceremonies conducted in fancy common spaces, it has served as a taste of Oxford or Cambridge life in the middle of New Jersey for more than a century. For Everett, it was his first opportunity to live miles from home and study in an institution neither military nor religious. As a bonus, he bonded with three other like-minded graduate students—Harvey Arnold, studying statistics, Hale Trotter, pursuing math, and Charles Misner, a budding young gravitational physicist straight out of Notre Dame and originally from Michigan. The group had a blast playing ping-pong and poker, as well as bouncing around weird ideas late into the evening over glasses of sherry.

"We were very close friends. . . . There was a group of four of us who got together. I was talking with [Everett] regularly," recalled Misner, who passed away in 2023 at the age of 91.[3]

Under John Wheeler's guidance, Misner was exploring ways of applying the techniques of quantum electrodynamics (QED)—the application of quantum principles to electromagnetic interactions—to a quantum treatment of gravitation. Wheeler's former student, Richard Feynman, had developed an extraordinary method used in QED called the path integral technique, informally known as "sum over histories."

Sum over histories is a kind of multiverse in a bottle, seen only when popping off its cork and recording the outcome. Imagine that two electrons emerge from different sources, approach each other, sense each other's electric repulsion, and scatter away from each other. Detectors measure the final result, known as the scattering profile.

Classically, those electrons would obey Newton's laws of mechanics, as well as Maxwell's laws of electromagnetism. Therefore, scientists could track their precise paths. However, following Heisenberg's uncertainty principle, quantum physics mandates a certain degree of haziness. No one could know both their precise positions and exact velocities at any given moment.

Feynman's innovation was to modify a classical notion, called the Principle of Least Action, to allow for an accounting of all of the possible paths taken by the electrons. "Action" is a tricky concept. A basic, nonmathematical explanation is that it encodes the energy efficiency of how bodies move through space subject to various forces. Least action, therefore, singles out the most efficient path—which happens to be the exact classical route through space. For example, a baseball thrown by a pitcher follows a sleek curve—a parabola—rather than a zigzag excursion through the air. The Principle of Least Action predicts such a smooth trajectory. Simply put, Feynman applied a quantum equivalent of action as a factor in a weighted sum of histories: a tally of all feasible trajectories from start to finish. While the least action path still represents the likeliest route, corresponding to the classical path, a haze of less probable trajectories surrounds it. Measurements of particle behaviors, such as scattering, must take into account the weighted sum of histories, not just the most probable traipse through time. Quantum actuality, therefore, is a select blend of possibilities: a casserole of alternatives carefully mixed together in just the right amounts, rather than just a single ingredient.

Image 11. Physicist John Wheeler, who spent much of his career as a professor at Princeton University, was a critical figure in the renaissance of general relativity. He mentored many students in that field and others. Credit: AIP Emilio Segrè Visual Archives.

Wheeler thought sum over histories was a brilliant idea, and taught it to his classes at Princeton. He ardently hoped it would help reconcile the dispute between his two intellectual heroes, Einstein and Bohr. Unfortunately, Einstein dismissed its probabilistic aspects as another form of "dice-rolling," and remained skeptical of its merits. Bohr attended a lecture by Feynman about it at the 1948 Pocono conference and was unimpressed for a different reason. The special sketches Feynman developed to represent various options for interactions—known, appropriately enough, as Feynman diagrams—rattled Bohr's conservative sensibilities about keeping quantum mechanisms a "black box" of unseen workings. Bohr mistakenly thought Feynman was trying to track actual movements, instead of capturing quantum uncertainty in a clever new way.

Despite the doubts of those great minds, Feynman's methods caught on in a big way. His diagrams and "sum over histories" technique are today ubiquitous in quantum field theory, which describes how subatomic entities interact by means of quantum rules, including QED and more advanced theories that model the nuclear forces. Curiously, while Feynman's approach includes alternative strands of reality, virtually no one objects to that kind of limited "multiverse." First of all, even hard-headed experimentalists recognize that it yields on-the-mark predictions of particle events. Secondly, those competing histories transpire in an abstract space; once real measurements occur, they are effectively blended together. The situation is in some ways analogous to the alternative worlds, according to the philosophy of Leibniz, confined to the mind of God; only the selected outcome becomes real.

Widespread acceptance of Feynman's approach points to the fact that any physicist urging that theories be confined to the observable must concede that quantum processes transpire in abstract domains of unlimited dimensions. Despite long-standing aspirations for realism among segments of the scientific community, no one has successfully constructed a truly open way of doing modern physics that eschews all manner of unseen realms and yet matches known experimental data.

Given the success of QED and sum over histories, Wheeler identified in Misner a mathematical prodigy potentially equipped to apply those methods to gravitation. Wheeler was overambitious by far, stressing to him that it would likely take around six months to complete the task. In fact, while Misner made inroads, resulting in his 1957 PhD thesis, "Outline of Feynman Quantization of General Relativity; Derivation of Field Equations; Vanishing of the Hamiltonian," along with many related publications, quantum gravity proved (and continues to prove) a tough nut to crack.

CONTENDING WITH BOHR

Increasingly, Everett's interests turned to solving the problem of quantum measurement theory. Sparked by his careful reading of von Neumann's textbook, in a class taught by eminent Hungarian physicist Eugene Wigner, and his discussions with fellow physics graduate students such as Misner, he sought a way of removing the need for conscious observation. Likely his attendance at Einstein's final lecture in April 1954—in which the sage, tongue-in-cheek, offered the curious example of a mouse serving as witness to a quantum measurement—also swayed him toward that research direction. Coincidentally, Wheeler was trying to build up his team by attracting more graduate students to the conundrums at the intersection of general relativity and quantum physics. Thus, Everett came to be a part of Wheeler's research group.

Yet another impetus for Everett's pursuit was a semester-long residency by Niels Bohr at the IAS in fall 1954. One highlight was a talk by Bohr on November 16 at the Graduate College on the philosophy of quantum mechanics, which Everett and Misner eagerly attended.

In advance of Bohr's stay, his research assistant Aage Petersen was spending time at Princeton. Loyal to his chief, Petersen loved to argue passionately about the validity of the Copenhagen interpretation. At the Graduate College, he wrangled with Everett (and to some extent Misner) on that topic. Everett relished the challenge of clanking swords with the Danish visitor.

As Misner recalled:

> Everett . . . was a brilliant odd-ball. His favorite sport was one-upmanship, which the Merriam-Webster dictionary defines as "the art or practice of outdoing or keeping one jump ahead of a friend or competitor." In the case of quantum mechanics Hugh's friend/competitor was Aage Peterson (an assistant to Niels Bohr who espoused the received wisdom known as the Copenhagen Interpretation of Quantum Mechanics). . . . Hugh thought it was preposterous to accept the Schrödinger equation as a fundamental law of quantum physics, but then assert that it is usually violated in any observation of quantum phenomena.[4]

Indeed, as he had with his theology professors in his undergraduate days, Everett loved poking holes in Petersen's reasoning. Misner watched the battle intently. He concurred with Everett that it appeared strange, for example, that the Schrödinger equation operated only part of the time. It functioned beautifully during the continuous phase of quantum evolution, but stopped working during the measurement process, switching to sudden wave function collapse instead. That certainly seemed a bizarre way for a law of nature to operate, they felt.[5] With his brashness amplified by the alcoholic concoctions running through his bloodstream, Everett mocked the orthodox position as being completely untenable. Surely a superior explanation must be found, he argued. Petersen would yield no ground.

As Everett recalled, when interviewed by Misner in 1977:

> One night at the Graduate College after a slosh or two of sherry . . . you [Misner] and Aage were starting to say some ridiculous things about the implications of Quantum Mechanics and I was having a little fun joshing you and telling you some

of the outrageous implications of what you said, and, ah, as we had a little more sherry and got a little more potted . . . in the conversation.[6]

Meanwhile, Wheeler was becoming increasingly immersed in the question of transforming general relativity and cosmology into quantum theories, thus helping complete theoretical physics. To do so, he believed, required assigning a wave function to the universe itself.

As Misner recalled, "Wheeler wanted to talk about the wave function of the universe. [He] wanted to make the wave function of the universe acceptable."[7]

The problem, though, with characterizing the universe in quantum terms, is who would be the observer? Clearly no human could stand outside the cosmos and induce its wave function to collapse. Nor could a mouse for that matter. Wheeler was at an impasse.

HOUSE OF MIRRORS

Wheeler was accustomed to unconventional ideas, having generated many of them himself. Once, when Feynman was a student, Wheeler phoned him and suggested that all of the electrons in the universe were the same particle, zigzagging forward and backward in time. From that strange notion, Feynman distilled the concept of treating positrons (electrons' positively charged antiparticle counterparts) as electrons going backward in time to aid in calculations. Such a shorthand has become a staple of quantum field theory, thanks to Feynman's clever reinterpretation of his mentor's offhand remark. Wheeler called his philosophy "radical conservatism": probing the frontiers of physics in an imaginative way while firmly keeping its fundamental laws in mind.

Despite his typical willingness to entertain bizarre concepts as long as they didn't blatantly violate natural principles, Wheeler was astonished when Everett approached him with a revolutionary way of removing observers from quantum physics. By making their conscious existence part of the continuous evolution of a "universal wave function," measurements would be seamless and operate without the need for human intervention. Nothing would collapse. Rather, the thread of conscious awareness would split upon each measurement, producing near-identical copies with one key difference. Each near-replica would record a different outcome. The universal wave function would encompass all of the near-clones. However, none would be aware of the others' existence. In short, instead of a house of cards facing collapse, quantum measurement would become a distorted house of mirrors, full of inexact copies.

Suppose a researcher wanted to measure the spin (response to an external magnetic field) of an electron using a magnet. Before the magnet is turned on, an electron's wave function is said to be in an equal mixture of two possibilities: spin-up (aligned with the magnetic field) or spin-down (opposite to the magnetic field). According to the Copenhagen interpretation, once she switches on the magnet, the wave function collapses to either spin-up or spin-down, with 50 percent likelihood for each. In Everett's model, the mixed quantum state persists, but, upon measurement, there are also two, near-replica copies of the observer. One records spin-up, the other spin-down, and nary the twain shall meet.

By analogy, let's imagine a theme park ride based on dragon stories. You board a car, and it winds through a dark, creepy forest. The car has a set of buttons that you can press at any time. Pushing one of them leads to a sword fight with a dragon. Pushing another results in a battle with that same dragon using a giant fire extinguisher. Yet another selection brings you into dialogue with that very dragon, who is able to speak and reason. The buttons aren't labeled however. Until you press a button, the car continues to meander around the forest. Once you do so, the track beneath the car diverts, sending it along a path toward the option selected.

Now suppose, in a moment of indecision or panic, you press all the buttons at once. The ride doesn't flinch. Magically, your identity divides painlessly into three near-replicas, each following a different track. The first slays the dragon, the second douses it, and the third convinces it to become nonviolent. Because all three versions of you are on separate tracks and never meet up, none find out about the others. Leaving via separate exits, the first "you" is so excited about slaying a dragon that it immediately leaves the theme park and enlists in the Marines. The second "you" joins a volunteer firefighter brigade, and the third, the Peace Corps. For the rest of your lives, you each think of yourself as the real "you." Moreover, from the ride operator's perspective, such splitting is simply a natural consequence of the tracking system, and everything has transpired normally and seamlessly. In fact, all three versions are aspects of the same happy customer. Such is the essential weirdness of Everett's idea.

Wheeler found Everett's idea unusual but intriguing. Given that, at the time, he couldn't think of a better way of characterizing the wave function of the universe, he offered to be Everett's supervisor and invited him to write up his idea as a PhD thesis project. Meanwhile Misner continued to work under Wheeler's supervision as well. Misner didn't believe in Everett's hypothesis, which seemed rather outlandish. However, he agreed that the Copenhagen interpretation was flawed and that alternatives were needed.

"I didn't like the conclusions, but I respected Hugh's ability to argue logically," noted Misner. "I certainly didn't like the Bohr hypothesis."[8]

Misner and Everett grew even closer after they decided to become roommates. In spring 1955, they both audited a course Wheeler taught in advanced quantum mechanics that centered on Feynman's sum over histories approach. At the same time, Everett continued to develop his idea. By the fall of that year, he started writing up his work, including a detailed report, complete with analogies along with mathematical analysis, called "Probability in Wave Mechanics." He handed his write-ups to Wheeler, who offered comments and suggested combining them into the rudiments of a thesis.[9]

As Everett's dissertation developed, Wheeler expressed both fascination and caution. He sincerely wanted to make Everett's idea work, so that the quantum umbrella could cover the universe as a whole without a need for outside observers to support it. However, to carry out such a rethinking of quantum measurement, he'd need to convince Bohr that it was at least a valid alternative. Wheeler's situation was akin, in some ways, to Copernicus's position when he dedicated *On the Revolutions of the Heavenly Orbs*, published in 1543, to Pope Paul III with the hope of making his argument for a sun-centered cosmology at least respectable enough to be considered a reasonable alternative to the geocentric canon. Not to elevate Everett to the status of a Copernicus, by any means. Rather, the latter's example shows that sometimes it is enough to establish a theory as falling within the domain of acceptable interpretation, rather than insisting that it usurp the canon and potentially generating far more resistance.

Bohr was one of Wheeler's heroes, supremely respected in the physics community (in some ways operating as the "Pope" of quantum research), and would make either a strong ally or formidable foe depending on the palatability of Everett's notion to him. Consequently, Wheeler expended much red ink insisting on cutting anything potentially controversial in the thesis. For example, Everett had come up with a clever analogy about intelligent amoebas splitting (the epigraph at the beginning of this chapter), which Wheeler urged him to cut—so he did. Wheeler wanted the focus to be on physical processes, rather than conscious experiences—the latter veering too much into psychology. Nevertheless, Wheeler's attempts to convince Bohr to at least consider the idea met with scant success. Bohr had hardened in his ways and had no interest in alternatives to the orthodox approach he had long advocated.

From January to September 1956, Wheeler was on sabbatical from Princeton, serving as Lorentz Visiting Professor at Leiden University, Netherlands. Misner and Everett parted ways, with the former deciding to join Wheeler for the rare opportunity to spend part of his graduate days in Europe, and the latter deciding to remain in the United States. He spent the spring at Princeton, continuing to work on his thesis, and the summer working at the Pentagon in an unrelated Department of Defense position.

During part of his stay, in May of that year, Wheeler headed up to Copenhagen to meet with Bohr. There, he tried to convince Bohr, as well as Petersen, of the merits of Everett's idea. He had hoped Everett could join them and make a strong case. Everett, however, had his hands full preparing his thesis and getting ready for his summer job. Either Wheeler was not very convincing, or Bohr was especially obstinate, or both, but his steel resistance wouldn't bend. Consequently, Wheeler, upon returning to Princeton in the fall, suggested even more extensive cuts to Everett's thesis, until it was virtually unrecognizable. By the time Everett's dissertation, originally "The Theory of the Universal Wave Function," was completed and defended in 1957 under the blander new title "On the Foundations of Quantum Mechanics," it was only one-quarter the size of the original manuscript. It would take more than a decade and a half until the earlier, clearer, and far-more-comprehensive draft was publicly available.

Given the radical cuts to his work and the cold reception by the mainstream physics community, especially Bohr, Everett was as discouraged as a minor-league baseball player rebuffed by major leaguers. Given such treatment, theoretical physics must have seemed to him more like a hobby than a profession. Naturally, he looked for a more viable career and decided to remain working for the Pentagon as a defense analyst. There, he helped develop the concept of mutually assured destruction in a nuclear war—such an unthinkable outcome that it arguably kept the Cold War from turning into an actual war with nuclear weapons. He eventually met Bohr during a trip to Copenhagen in 1959, but once again with nothing to show for his efforts.

TOUCHÉ, YOUNG PHYSICIST!

In January 1957, a formerly obscure branch of physics received a much-needed jump start when French-American mathematical physicist Cécile DeWitt-Morette organized the first American conference on general relativity and gravitation. Held in Chapel Hill, where she and her husband, American gravitational physicist Bryce DeWitt, were visiting research professors (and the latter, the director of the Institute of Field Physics) at the University of North Carolina, the conference "On the Role of Gravitation in Physics" attracted eminent researchers from around the world including Feynman, Wheeler, Bergmann, and many others.

Given Einstein's massive fame, one might wonder how research in general relativity and gravitation had been little known prior to the Chapel Hill conference—save a few enclaves, such as Wheeler's group at Princeton. The main issue was that two of Einstein's key predictions—the rate of Mercury's orbital precession and the bending of starlight by the sun as seen during a total solar eclipse—had already been verified. The simplest solutions to his equations, such as Karl Schwarzschild's "black hole" model and several different models of the dynamics of the universe, starting with Einstein's own cosmology, had been found. Moreover, Einstein spent the final three decades of his life searching for a unified theory of the natural forces that superseded general relativity. He effectively declared general relativity to be incomplete. With its founder having moved on, and seemingly very few projects left to complete, theorists didn't want to dip into a shrinking puddle—especially when there was a whole sea of projects to explore in thriving areas such as nuclear and particle physics.

There was a direct line between Wheeler's interest in and advocacy of general relativity at Princeton, and its blossoming at the Chapel Hill conference. Fresh off working on the hydrogen bomb project, Wheeler had a pragmatic side and encouraged military and industrial support of gravitational research. Industrialist Agnew H. Bahnson Jr. of the Bahnson Company, an air-conditioning company based in Winston-Salem, North Carolina, would play a key funding role. The hope was to "harness gravity" to provide propulsive power—for peaceful use or warfare. Bahnson's support helped lead to the establishment of the Institute of Field Physics, the appointment of Bryce DeWitt as its director, and the funding of the conference.

Image 12. Theoretical physicist Bryce DeWitt, who coined the term "Many Worlds," popularized Everett's theory, and pursued the quantization of gravity, among his numerous contributions. Credit: AIP Emilio Segrè Visual Archives, Physics Today Collection.

The Chapel Hill conference spurred ample discussion about ways to quantize gravitation, the major thrust of DeWitt's research at the time. There was also much discourse about the then-speculative (and now confirmed) notion of gravitational waves: ripples in the fabric of space-time caused by gravitational disturbances such as collisions between massive bodies. Wheeler's group at Princeton was well represented, including his students Misner and Dieter Brill, as well as visiting researchers Joseph Weber and Edwin Power.

While Everett didn't participate in the conference, Wheeler sought to get the word out about his innovative dissertation work. A respected journal, *Reviews of Modern Physics*, dedicated a special issue to Chapel Hill, edited by DeWitt. As DeWitt was in the process of collecting papers by conference presenters to publish in that volume, he noticed that Wheeler had sent him a paper by Everett, titled "Relative State Formulation of Quantum Mechanics." In the same envelope was a compelling cover letter by Wheeler, as well as an interpretative article by him, each explaining the importance of Everett's work. Given Wheeler's prominence and support for his work, and the connection of Everett's research to the themes of the conference, DeWitt gladly agreed to include Everett's and Wheeler's papers.

In his accompanying piece, Wheeler emphasized the radical nature of Everett's construct. He compared its potential impact on science to that of the revolutionary works of Newton, Maxwell, and Einstein, writing:

> It is difficult to make clear how decisively the "relative state" formulation drops classical concepts. One's initial unhappiness at this step can be matched but few times in history: when Newton described gravity by anything so preposterous as action at a distance; when Maxwell described anything as natural as action at a distance in terms as unnatural as field theory; when Einstein denied a privileged character to any coordinate system, and the whole foundations of physical measurement at first sight seemed to collapse.[10]

In reading over Everett's contribution, DeWitt was floored by its implications, which, at first, seemed as contrary to our common experience as purporting that people float in the air. Standard quantum interpretations such as Heisenberg's, he noted, emphasize the transition from possibility to actuality through the process of measurement. If wave functions don't collapse and we find ourselves in an evolving mixture of quantum possibilities, doesn't that mean that things get weirder and weirder over time?

"I still recall vividly the shock I experienced on first encountering this multiworld concept," remembered DeWitt. "The idea of 10^{100} slightly imperfect copies of oneself all constantly spitting into further copies, which ultimately become unrecognizable, is not easy to reconcile with common sense."[11]

Consider, for example, Schrödinger's cat conundrum, a thought experiment in which a scientist places a hapless feline within a closed box, along with a radioactive sample, a Geiger counter, and a trigger mechanism that releases poison if the sample is recorded as emitting a decay particle within a given time interval—which has a 50 percent chance of occurring. While the sample's wave function remains in a quantum superposition of not-decayed and decayed, the cat's wave function persists (or "purr-sists" perhaps?) in a zombie-like blend of alive and dead. Applying the Copenhagen interpretation literally—and, as Schrödinger thought, absurdly—only when the scientist opens the box would the cat's wave function collapse into one of the two options: spared or deceased.

If that isn't strange enough, Everett's interpretation, in contrast, offers the bizarre outcome that the cat's state would continue to stay blended even after the box is opened. In tandem with that superposition of possibilities, the scientist's state would seamlessly bifurcate. One near-replica would be relieved to witness a living feline and the other heartbroken to encounter the lifeless victim of a cruel experiment. Furthermore, according to Everett's hypothesis, such branching of reality would be happening almost continuously at a staggering rate.

Surely, being near-replicated, DeWitt supposed, is not the kind of process one pictures as part of a normal day. Brushing one's teeth, combing one's hair, buttoning one's shirt, and doubling one's conscious existence—one of those things doesn't quite belong. Science should be grounded in real-life experiences, he felt, not in fanciful musings about alternative worlds.

DeWitt penned a detailed response to Wheeler, balancing his deep curiosity about Everett's notion with his hesitancy to believe in its physical implications. Craft any theory you like, but its true measure, most clear-eyed physicists believe, lies in its experimental verification. Given that no one would ever attest to splitting like an amoeba, who could serve as witness to Everett's predicted multiplicity of outcomes?

"The trajectory of the memory configuration of a real physical observer," DeWitt wrote, "does not branch. I can testify to this from personal introspection, as can you. I simply do not branch."

After Wheeler shared DeWitt's critique with him, Everett decided to write back directly and defend himself. He cleverly drew an analogy with critiques of Copernicus's notion that Earth rotates around its axis and revolves around the sun. As Everett wrote:

> One of the basic criticisms leveled against the Copernican theory was that "the mobility of the Earth as a real physical fact is incompatible with the common-sense interpretation of nature." In other words, as any fool can plainly see the Earth doesn't really move because we don't experience any motion. However, a theory which involves the motion of the Earth is not difficult to swallow if it is a complete enough theory that one can also deduce that no motion will be felt by the Earth's inhabitants (as was possible with Newtonian physics). Thus, in order to decide whether or not a theory contradicts our experience, it is necessary to see what the theory itself predicts our experience will be. . . .
>
> I can't resist asking: Do you feel the motion of the Earth?[12]

DeWitt was extremely impressed by the quality of Everett's reasoning. "His reference to the anti-Copernicans left me with nothing to say but 'Touché!'" noted DeWitt. "His reply to my letter was succinct and to the point. I had no further ammunition to throw at him."[13]

THE EVER-GROWING TREE OF BRANCHING REALITIES

In the decade following the Chapel Hill conference, while Everett worked as an analyst in the Defense Department and largely lost touch with the physics community, DeWitt remained at the University of North Carolina continuing to pursue a quantum theory of gravitation.

He continued to be impressed by Everett's interpretation. Given that the universe has no external observers, he noted that a full quantum description, including a "universal wave function," must operate smoothly without the need for measurement. Everett's hypothesis, including observers as part of the grand description, rather than as external elements, seemed to fit the bill. Therefore, DeWitt concluded:

> There is no way to interpret the wave function of the universe except for Everett.[14]

In the late 1960s, DeWitt was astonished and disappointed to learn that Max Jammer, a respected physicist and philosopher of physics who was working on a treatise about interpretations of quantum mechanics, had not even heard of Everett. As DeWitt recalled:

> I thought that this was scandalous, because Everett had a brand-new idea, it was the first fresh idea in quantum theory in decades, and . . . he was being completely ignored. So, I decided to write . . . a popular article, for *Physics Today*, which really put Everett on the map, and Wheeler promptly disowned Everett. The reason, as far as I can see, was it was too revolutionary an idea, this idea of many worlds. It was anti-Copenhagen; Bohr was one of Wheeler's heroes, and he didn't want to be associated with him. He has denied, he has refused to have anything to do with it in all the years since.

DeWitt's article in *Physics Today*, a journal read by members of the general public as well as career physicists, truly put Everett's idea on the map. With the catchy title and subtitle "Quantum Mechanics and Reality: Could the Solution to the Dilemma of Indeterminism Be a Universe in Which All Possible Outcomes of an Experiment Actually Occur?" the piece demonstrated his gift for framing physics research in a compelling way. In subsequent talks and publications, DeWitt decided, for the purpose of clarity, to coin an evocative expression, the "Many-Universes Interpretation of Quantum Mechanics," which became more familiarly known as the "Many Worlds Interpretation (MWI)."

DeWitt had previously called the idea the "Everett-Wheeler interpretation," assigning Wheeler partial credit for sponsoring and advocating for it. Then, shortly after his *Physics Today* piece appeared in September 1970, he noticed that Wheeler seemed to disavow the idea. DeWitt speculated that it was his loyalty to Bohr—who died in 1962—kicking in. However, it is unclear why, if Wheeler defended Everett's idea—albeit weakly—when Bohr was alive, he'd be even more cautious after Bohr's death.

More likely, Wheeler was starting to ruminate about imaginative ways to collapse the universal wave function by means of conscious observation. That path would eventually lead him to the "Participatory Universe" idea, in which the observation of the early universe by present-day astronomers and other individuals leads retroactively to the collapse of the quantum state in the past. That collapse would need to proceed in such a way that leads to living planets and conscious entities. Hence, the entire process would be a multi-billion-year feedback loop: consciousness inducing collapse and collapse (eventually) spawning consciousness. His concept was deeply connected with the Anthropic Principle, a universal selection concept developed, as we'll discuss, by Australian-born physicist Brandon Carter.

In 1972 the DeWitts moved to the University of Texas at Austin, where each was appointed to a professorship. Several years later, upon Wheeler's retirement from Princeton, they suggested that he join them in the same physics department for a post-retirement position. Wheeler agreed, relishing the opportunity for frontier research. His efforts came to focus on speculative theories that information content, such as arrangements of computer bits, is fundamental to understanding the universe. Somehow, he believed, that notion, which he called "it from bit," would mesh with his Participatory Universe idea in a kind of metaphysical rumination that he would later call "How come the quantum?" In that self-referential scheme, he managed, somehow, to come up with a description of nature even stranger, in some ways, than Everett's.

In 1977, Wheeler and Bryce DeWitt reached out to Everett and invited him to give a talk to their department. The theme Wheeler suggested, "human consciousness and the problem of a computer's 'consciousness,'" connected with his own research focus. Although he didn't enjoy public speaking and disagreed with Wheeler's novel speculations, Everett was grateful for the newfound interest in his work and gladly agreed to deliver a lecture. The one "perk" he asked for was the right to smoke in the lecture hall. In those days when public smoking was more common, the organizers agreed.

One of the attendees at Everett's lecture was young graduate student David Deutsch, who was engaged in fundamental research at UT Austin—officially under the guidance of Wheeler and effectively also under DeWitt. He would later go on to be a pioneer in quantum computation and other fields, winning many awards and accolades—most recently sharing the 2023 Breakthrough Prize in Fundamental Physics. During lunch in a "beer-garden restaurant," DeWitt made sure that Deutsch was seated next to Everett, so the two could chat about Everett's ideas, including Many Worlds.

Deutsch, who is now a professor at Oxford, recalled that Everett was a "very impressive person—full of nervous energy, highly-strung, a chain-smoker, very much in tune with the issues of the interpretation of quantum mechanics, unusual for one having left academic life many years before."[15]

After Everett's talk, Wheeler started to ponder ways of getting him involved again in theoretical physics. In 1979, he tried to convince a prestigious research center, the recently established Institute for Theoretical Physics associated with the University of California, Santa Barbara (now the Kavli Institute for Theoretical Physics, or KITP), to set up a quantum measurement unit and hire Everett as a visiting scholar, at the very least, but that recruitment never came to pass. Sadly, Everett would die in 1982, at the age of only fifty-one, of a sudden heart-attack. His health had been poor due to a combination of chain-smoking, heavy drinking, and obesity.

Everett left behind a wife, Nancy, a daughter, Liz, and a son, Mark. Afflicted with mental illness, Liz took her own life in 1996, writing in a suicide note that she wished to join her dad in a parallel universe. Mark, a musician, founded the rock band the Eels as its lead singer and songwriter under the stage name "E." In the 2007 documentary *Parallel Worlds, Parallel Lives*, he brought his father's life story to greater attention, rekindling the public interest generated by DeWitt's popularization decades earlier. Though Everett's far-reaching notions about the nature of reality and bold confrontation with the quantum establishment still resonate with many philosophically minded thinkers, along with fans of popular science, his hypothesis has yet to enter standard canon—fittingly, many hard-headed physicists would say, due to its lack of experimental evidence.

PHILOSOPHICAL BAGGAGE CLAIMS

Despite DeWitt's cogent advocacy and a surge of interest that followed, the Many Worlds Interpretation has remained a minority view in the mainstream scientific community—for example, as gauged by its meager or nonexistent treatment in introductory quantum mechanics textbooks. Most primers continue to refer to the collapse of the wave function as an indelible part of quantum theory. If the MWI is mentioned at all in a textbook, it is typically lumped near the end along with an assortment of speculative quantum alternatives.

Many practical-minded physicists, accustomed to follow the idea espoused by William of Ockham (widely known as "Occam's razor") to choose the simplest explanation whenever possible, have rejected the MWI as carrying too much philosophical baggage. For example, eminent theorist Freeman Dyson wrote:

I do not remember when I first heard of the Everett interpretation. I always disliked it and considered it a stupid waste of time to discuss it. Borrowing the words of Pauli, I would say that it was "not even wrong."[16]

Science writer Philip Ball, a prominent MWI critic, has similarly argued that researchers should focus on testable, rigorous mechanisms, rather than fanciful notions reminiscent of science fiction. As he wrote:

We should resist not just because MWI is unlikely to be true, or even because, since no one knows how to test it, the idea is perhaps not truly scientific at all. Those are valid criticisms, but the main reason we should hold out is that it is incoherent, both philosophically and logically. There could be no better contender for . . . Pauli's famous put-down.[17]

To rebut such scathing criticism, many MWI proponents argue that the essential weirdness lies in quantum physics itself—with its ever-changing superposition of states housed within in a mind-bogglingly intricate Hilbert space—not in its interpretation as branching realities. The real excess philosophical baggage, they assert, stems from taking continuously evolving quantum mechanics and tacking on a completely different mechanism to collapse the wave function. In comparing the MWI to the Copenhagen interpretation, they recount Everett's views on how unnatural it is to introduce human intervention into physical processes. Physics is simpler, they contend, by letting quantum states evolve on their own, and accepting the implications no matter how unintuitive. In other words, quantum mechanics narrates its own story, and that is the MWI.

Yet the original conceit of the Copenhagen interpretation that the world of humans stands completely separate from the realm of quantum experiments no longer strictly holds. Scientific developments in the past half-century have broken down the artificial wall between the measurer and what is measured. Consequently, MWI advocates no longer can contend that theirs is the only theory that bridges the gap.

In the early 1970s, innovative German physicist H. Dieter Zeh shed considerable light on the issue of quantum measurement with his discovery of a phenomenon called decoherence. Agreeing with Everett's position that observers and the observed shouldn't be separated descriptively into distinct realms with different rules, he examined how exposure of a quantum system to the environment—including measuring devices—constitutes a type of quantum entanglement.

Quantum entanglement—a phenomenon that Einstein derided as "spooky action at a distance"— is the idea that quantum information can be shared nonlocally. In the early history of quantum mechanics, theorists such as de Broglie and David Bohm tried to dispute it by means of "hidden variable" theories that re-establish locality and determinism in nature through unseen mechanisms. However, after Northern Irish physicist John Stewart Bell, in 1964, developed an important litmus test for determining whether or not quantum information could be shared via continuous local means—a back channel, so to speak—all known quantum entanglement experiments have ruled out hidden-variable explanations. Three physicists conducting quantum entanglement and teleportation experiments—Alain Aspect, John Clauser, and Anton Zeilinger—shared the 2022 Nobel Prize in Physics for their meticulous findings showing nonlocal effects that cannot classically be explained. Without a doubt, quantum entanglement is real.

According to decoherence theory, because of quantum entanglement, certain orderly patterns that the original system embodied—such as the lining up of phases (quantum pointers oriented in various compass directions)—might become "polluted" by the environment and irrevocably transform into disorderly states. Picture a sealed case of fancy watches with all of their second hands synchronized, opened and distributed to construction workers working with riveting equipment at a building site. The jolting environment might well bring the watches out of sync. Similarly, Zeh found, environmental interactions might disrupt the coherence of a formerly isolated quantum system. Experiments have confirmed that quantum systems, once exposed to the environment, do indeed appear to degrade.

Zeh's revelation of a natural, irreversible connection between quantum systems and external elements such as measuring devices represented a milestone in helping bridge the divide between the subatomic and mundane realms. While it didn't fully explain the process of metamorphosis into classical results, it effectively burst open Bohr's "black box" approach, and showed how von Neumann's divide between continuous evolution and instantaneous transformation was simplistic. Finally, it brought quantum processes in line with the Second Law of Thermodynamics and its notion that disorder doesn't naturally decrease and often tends to increase—resulting in a unidirectional arrow of time.

One of the knotty issues with quantum physics that decoherence theory strives to untangle is the "preferred basis problem." Classically, when we use precise tools to measure the location of a particle we expect to see a definite position, not a blur of options. Life and death are similar absolutes. Zoo exhibits normally contain living animals and many natural history museums include dioramas with deceased animal specimens, but—horror films aside—we'd be astonished to encounter living-dead blends.

The Copenhagen interpretation—linked with Bohr's notion of complementarity—ensures definitive outcomes by inserting the experimenter's choice of apparatus into the equation. The decision to use an apparatus that tests location, it asserts, results in the system assuming a position state basis. Cruelly, the poison release mechanism in Schrödinger's cat thought experiment establishes that one of its two basis states represents life and the other death. Therefore, the opening of the box results in a definitive outcome.

Decoherence theory removes the experimenter's volition from the picture, and posits that, once the quantum system is exposed, something in the environment selects the basis, and causes irreversible degradation of the original mixed state into one of the basis states. Naturally, the apparatus could play a major role in creating bias toward a particular basis—for example, a magnet forcing a superposition of electron spin states into one of two choices: spin-up or spin-down. The human brain might also help guarantee definitive states if its network of neurons offers a hostile environment for coherent quantum superposition and instead forces ensembles of states to decohere into one of the focused options.

Despite the decoherence breakthrough, the quantum measurement enigma remains far from resolved. Theorists continue to debate the question of how quantum processes, with their built-in ambiguity, transform reliably and completely into the definitive outcomes we witness in everyday life. The way forward could well include the MWI or, alternatively, completely eschew it in favor of a more tangible explanation.

As quantum theorist Wojciech Zurek has noted, "Decoherence is of use within the framework of either of the two major interpretations: It can supply a definition of the branches in Everett's many-worlds interpretation, but it can also delineate the border that is so central to Bohr's point of view."[18]

A fully satisfying theory of quantum measurement would need to explain why, in measuring observables using proper detectors, we absolutely never see blended states engaged in the process of settling down into one of the options. Rather, we reliably detect single outcomes. Decoherence takes us far, but does not quite demonstrate why, under ordinary circumstances, we always witness classical behavior, consistent with Newton's laws of mechanics, and never catch even fleeting glimpses of quantum blurring.

To guarantee that in taking precise quantum measurements, we consistently perceive definitive results, as in the classical world, there are a number of schools of thought. MWI purists remain confident that a medley of near-replica observers ensures that each would record a unique result. Each would have the illusion of living in a classical domain, unaware that they are part of an evolving quantum superposition. A variation of that idea, known as the "Many Minds" interpretation, envisions entanglement between an observed quantum system and the brain of its observer. A single observer's brain would host a medley of different mental states corresponding to the distinct measurements. On the more practical side are various schemes for "spontaneous localization" that use amplified random nudges from the environment to cause wave functions to collapse rapidly down to definitive states without the need for conscious observers. In short, quantum measurement hypotheses range from purely mental to completely physical.

WITHIN YOU, WITHOUT YOU

None of the quantum measurement options are perfect. One compelling criticism of the MWI is that it doesn't naturally explain what is called the "Born rule." Proposed by Max Born in 1926, it represents a way of making predictions in quantum mechanics. For each observable—position, momentum, spin, or otherwise—theorists can determine the probability of particular outcomes by expressing a system's quantum state in its appropriate basis and calculating the weightings of its eigenstate components. Mathematically, the probabilities turn out to be the squares of those weightings.

In any quantum measurement scheme involving collapse, such predictions can be tested by conducting multiple trials and recording the frequency of each outcome. For example, based on the wave function of an electron trapped in a box that is one centimeter wide, one might compute the chances of it being less than one millimeter away from the left side of the box. Suppose those calculated odds, let's say, happen to be 10 percent. That means that in one out of ten position experiments, set up under the same conditions, the electron would be found within that one-millimeter range.

However, in the MWI, there are two major obstacles to making sure that the Born rule is followed. The first is that the preferred basis problem lacks a natural solution. Nothing ensures that, for a position measurement, reality splits along the lines of the components of the position basis, rather than, say, the momentum basis. Second, even if the splitting somehow does occur according to the proper basis components, there is no practical way of testing that the chances of being in a certain branch match the Born rule predictions.

If there are 50-50 odds, for instance, such as in Schrödinger's cat conundrum, and the universe bifurcates into realities with two different outcomes, we would have to imagine near-replica observers with contrary experiences and memories. Neither would be able to attest that they had a single near-duplicate who witnessed the opposite result, hence confirming the Born rule's prediction of equal chances. For all they knew, there could be an army of near-clones who experienced the alternative. They couldn't directly check.

Though weird to envision such a splitting, at least it is a straightforward near-duplication, like cell division. The situation changes with quantum processes that involve a continuous distribution of probabilities representing a wide spectrum of possible outcomes. Simple branch-counting—enumerating how many branches end up with each outcome—would not necessarily yield the correct odds. Pure bifurcation into two equally likely branches would correspond to 50-50 odds, for example. Yet if you were measuring that electron in a box, you would expect one out of ten copies of you measuring it to be within the one-millimeter distance range, and nine out of ten copies recording otherwise. Now, if the odds of something happening, as calculated by the Born rule, turn out to be one in a trillion, picturing a trillion copies of an observer, with all except one in agreement, seems rather contrived. That would happen for every single quantum measurement throughout history. How might we establish such a complex pattern of replication given the lack of communication between its byproduct worlds?

Resolving the thorny issues involving applying the Born rule to the MWI has become an ongoing challenge for physicists and philosophers studying quantum measurement theory. One line of reasoning, initiated by Deutsch, and further developed by University of Pittsburgh philosopher of science David Wallace (formerly at Oxford, where his work on the topic began), involves applying the science of decision theory to assign weighting to different branches. It posits that rational decision-makers with knowledge of the workings of physics and invited to bet on the outcome of an experiment would reliably offer predictions consistent with the probability distribution delineated by the Born rule. Thus, one needn't run an experiment repeatedly, record the frequencies of various outcomes and match probabilities to those frequencies to establish that the Born rule holds. Rather, an ensemble of rational thinkers asked to make meaningful wagers on the results of an experiment, with the goal of being rewarded later if correct, would reliably match the probability distribution set out by the Born rule without reference to chance and frequency.

In addressing the preferred basis dilemma, Wallace has crafted a creative proposed solution based on the philosophical concept of emergence. Emergence, in philosophy, is when large-scale features arise from the seeming pandemonium of myriad microscopic components, none of which possess those properties. For instance, an artist might decide to render the Mona Lisa on a concrete wall with spray paint, on a beach with colored sand, or on a computer screen with pixels. In each case, presuming the artist were competent, we'd immediately recognize Leonardo's masterpiece in the big picture—but likely never guess it by analyzing the various components themselves. Similarly, Wallace has posited that quantum states take on various bases and divide into various worlds in a hodgepodge of ways. It is only through emergence that we perceive classical, Newtonian trajectories from that jumble of evolving blended states.

Wallace draws on the animal kingdom for a clever analogy. "Consider tigers," he writes, "which are (I take it!) unquestionably real, objective physical objects, even though the Standard model contains quarks, electrons and the like, but no tigers. Instead, tigers should be understood as patterns, or structures, within the states of that microphysical theory."[19]

Accordingly, Wallace argues, while the quantum realm is real, independent, and diverse, it would be impossible to fathom all of its components. Rather, our experiences steer us toward studying emergent features, such as likeliest paths through space, that offer the continuity we expect from the tangible world.

Deploying somewhat different tactics to try to solve the puzzles of the preferred basis problem and the Born rule for the MWI, Oxford philosopher Simon Saunders has drawn upon decoherence theory as well as techniques derived from the nineteenth-century statistical methods of Ludwig Boltzmann. Decoherence theory is powerful enough, he argues, to create a branching structure that aligns with the basis expected for particular kinds of experiments, such as a quantum spin basis—some branches with spin-up and other branches with spin-down—for an electron tested for its reaction to a magnetic field. At that point, following Boltzmann's way of counting the number of microstates (types of behavior on the atomic scale) that meet certain qualifications, such as average kinetic energy, to create a macrostate (the set of mundane characteristics, such as temperature and pressure), Saunders develops a way of branch-counting that lumps together branches—the microstates—associated with similar overall properties—the macrostate. The result is a statistical way of replicating the Born rule.[20]

Until a complete model of conscious awareness emerges in neuroscience, however, the ultimate step of quantum measurement—mentally recording particular outcomes within the context of subjective experiences of reality—will remain a mystery. Both the Copenhagen and MWI interpretations share an Achilles' heel in their treatments of how the mind works. In the former, conscious observation triggers collapse, and in the latter, measurement results in a splitting of conscious observers into near-replicas. Their conscious awareness is thereby shared by multiple versions of the same person in different parallel realities, each of whom thinks he or she is the "real one." The problem is, as pointed out by many theorists, we really do not understand consciousness. Given that it is tricky to define self-awareness, except to say that everyone knows they have it and thereby presumes others do too, it remains unclear what is the exact relationship of minds (the conduits of free will) to the physical world.

Human brains are made of atoms, though, which, according to decoherence theory, would become entangled with what is being measured. As a result, there would be no obvious reason for measurers to experience distinct outcomes, rather than a mélange of states. In other words, rather than being independent observers of the separate options recorded in experiments, they'd be physically enmeshed with the quantum vagueness of the wave function and sense only a blur.

To rectify that situation, H. Dieter Zeh proposed what has become known as the "Many Minds" interpretation. Making use of the fact that science still doesn't fully know how conscious existence is tied to brain function, it allowed for Everett's branching to take place in the form of near-replica minds, rather than near-replica people. Consequently, the classical notion of a singular measurement rather than a medley of possibilities would become a mental event, similar to waking up after a dream and realizing, retroactively, what it was about. Given that minds are constantly weeding out noise when faced with a deluge of sensory information, perhaps parallel mental states, each with unambiguous results, are a way of processing quantum mayhem.

In contrast to Everettian proposals, such as the MWI and Many Minds, a number of teams of researchers have attempted to extend decoherence theory by developing physical models of how quantum states collapse. Proposed schemes for "spontaneous localization" rely on a quantum system's physical surroundings to trigger collapse down to unequivocal outcomes. The basic idea is that environmental interaction, assisted by a hypothetical amplification mechanism that turns light pokes into heavy jolts, would rapidly induce the dynamic reduction of a quantum state from a blend of possibilities down to a single option, without the need for observers. The first such proposal, "Unified Dynamics for Microscopic and Macroscopic Systems," based on the buildup of random tugs between particles and their surroundings, was published in 1986 by Italian physicists Giancarlo Ghirardi, Alberto Rimini, and Tullio Weber. It was followed by a related proposal introduced by American physicist Philip Pearle in 1989 and further developed with the help of Ghirardi and Rimini. Yet another method, relying on minute gravitational effects to initiate the collapse, was proposed by Lajos Diósi the same year and independently suggested by Roger Penrose in 1996.

Most reduction mechanisms focus on localizing position and other parameters present in classical, as well as quantum, physics. But what about exclusively quantum phenomena such as changes in polarization and spin states due to alterations in what is being measured? Explaining all such behavior using dynamic reduction mechanisms would be a tall order. A clear advantage of MWI is its flexibility in allowing for every viable alternative as a distinct branch of a universal wave function. No physical mechanism for collapse needs to be tweaked, because there is no collapse.

Without a doubt, modern physics is faced with a dilemma. The conventional world offers precise results on a regular basis. Press a button on an elevator panel, and it will arrive at a particular floor. Roll a bowling ball, and it will either knock down a certain number of pins or land in the gutter, largely depending on its initial speed and direction. Even many things that seem random have predictable outcomes. Spin a roulette wheel and it will land on a particular number—which might be anticipated given its initial state, rate by which it is set into motion, operating conditions, and so forth.

The quantum world, on the other hand, is akin to Hilbert's hotel in possessing a medley of occupied rooms hidden behind closed doors. Somehow, upon inquiry, one of those doors seemingly opens and its occupant—a single result—emerges. Could it be that there is a dynamic reason for one, and only one, such portal to spring open? Or is it the process of conscious observation that leads to that result?

Alternatively, do near-duplicate versions of ourselves—or perhaps just our minds—visit every single one of a range of such rooms in Hilbert's quantum hotel? In that grand open house, we have unlimited access but just don't realize it, given the limitations of our conscious experience. If so, to rework a famous Shakespeare passage, how replicative mankind is in our brave new multiverse!

Many of those who think little of Everett's proposal continue to hold out hope for some form of spontaneous localization or other credible mechanisms for collapse. At the very least, he deserves credit for rankling the establishment and carving out space for opposing ideas to the Copenhagen interpretation that aren't simply fruitless searches for hidden variables.

Aside from addressing the tricky issue of quantum measurement, there are many other reasons to entertain multiverse ideas. In explaining why the observed universe has the particularly favorable conditions it does—leading to stable stars, habitable planets, and ultimately sentient beings such as ourselves, it is instructive to consider the alternatives. One way of addressing such alternate possibilities is to embrace the notion of a multiverse of vying realities, with some fitter for life than others. More cautiously, one might strive to find dynamic reasons for the universe's current state, such as a mechanism connected with the properties of general relativity and its cosmological solutions. Parsimony suggests attempting means within our own universe, before taking the conceptual leap of positing others.

In fact, during the late 1960s theorists such as Misner ardently attempted to find a solid reason why—amongst all possible cosmological solutions to general relativity—our universe ended up so regular.

Focusing on a particular model that he dubbed the "Mixmaster universe," he hoped to show how even utter chaos at the start of the universe would smooth itself out and lead to the evenly distributed sky conditions we observe today. Physicist Robert H. "Bob" Dicke, another Princeton maverick, tried to modify general relativity itself, replacing it with a theory with a changing gravitational strength parameter, with a similar goal of establishing why the universe currently is the way it is. However, Dicke also recognized the power of invoking our own existence as a selection mechanism for weeding out less desirable possibilities. In the hands of Brandon Carter, that idea became known as the Anthropic Principle. One version, the Strong Anthropic Principle, relies on selection within a multiverse of competitors. Carter acknowledged that he had the MWI in mind while formulating that idea. For him, invoking the notion of a multiverse is a perfectly reasonable "plan B," when dynamic attempts pertaining to a single universe (the Mixmaster model, for example) fall short.

Virtually all physicists would agree that dynamic methods, based on time-tested physical laws and mathematical relationships, lead to superior explanations of natural phenomena. Where there is considerable controversy is when (if ever) to invoke speculative reasoning based on the hypothesis of a multiverse of alternatives. The other options might be part of a "dreamscape" of imagined alternate paths, or a fully realized "landscape" involving actual parallel universes that are "out there" somewhere. We employ something like the former in our daily lives whenever we ruminate about roads not taken. For instance, we might think about scenarios in which our parents never met, and categorize them as hypothetical worlds that were unrealized. No one seriously tries to construct a dynamic method based on the laws of physics to justify how the molecules that led to their own embryo happened to be in that particular configuration. That would be some weird way of explaining the "birds and the bees." Is it such a stretch, some argue, to employ similar existence arguments involving selection from a range of alternatives when dynamic explanations seem impossible?

On the other hand, for a segment of contemporary physicists, anthropic reasoning and reference to a multiverse are simply cop-outs—reflecting a current dearth of innovative solutions to physical problems. In that view, such methods hark back to a time before the scientific method, when philosophers would simply reason why there must be a certain number of elements in the universe, for instance. Why not wait, such multiverse objectors argue, until objectively testable new theories emerge? Indeed, if such novel approaches prove to be just around the corner, they might well be right. But if they aren't forthcoming for decades, calls for a more conceptual narrowing-down—by means of selection within a multiverse, or otherwise—will only grow louder.

Cogito ergo mundus talis est (I think therefore the world is such).

—Brandon Carter, "Large Number Coincidences and the Anthropic Principle in Cosmology"

CHAPTER FOUR

ORDER FROM CHAOS

Charles Misner's Mixmaster Model Versus Brandon Carter's Anthropic Principle

As a respected theorist with a solid publication record in serious journals, Bryce DeWitt's vocal advocacy of the Many Worlds Interpretation undoubtedly affected the culture of physics. References to multiple universes increasingly became a rhetorical tool to place our own cosmic conditions in context. Along the lines of a thesis by science historian David Kaiser, one might speculate that the popular culture of the time—the hippie movement and the Summer of Love—also played a role in tuning in minds to multiverse ideas.[1] Just as psychedelic experiences opened participants' eyes to realms beyond ordinary life, could it be that our staid, observable universe is embedded in the equivalent of a Ken Russell film of cosmic madness? The "chaotic cosmology programme"—an expression coined by John Barrow in the 1980s, but apt even for the late 1960s and early 1970s—examined how, out of all the possible sets of physical constants and types of cosmic dynamics, leading to myriad weird alternatives, our own slow, steady, and evenly distributed spatial expansion developed in a way that allowed us to be here.

In modern physics, if one includes purely mathematical models, envisioning alternative realities is as basic as breathing. Even without introducing extra dimensions, such as in Kaluza-Klein theories, or invoking quantization, such as in the hypothesized "universal wave function," Einstein's general theory of relativity in its standard, classical form possesses an extraordinarily rich range of solutions. Like a toy chest in an enriched kindergarten classroom it houses building blocks of innumerable shapes and sizes. Some are positively curved, akin to the surfaces of spheres. Others are negatively curved, like saddle shapes. Yet others are flat, but in three dimensions, like an infinite box. That cornucopia of conceivable geometries satisfies versions of general relativity either with or without the cosmological constant—a stabilizing term added and later rejected by Einstein. Narrowing that cacophony of solutions down to the humdrum regularity of the physical universe we witness today posed a daunting challenge.

Einstein, in his own cosmological explanations, initially hoped that only a single model of the universe would prove self-consistent and match all known observations. The other mathematical solutions, he believed, could then simply be rejected as unphysical curiosities, akin to imaginary numbers. Even when the gamut of mathematically viable solutions was brought to his attention, he stuck to the notion of singling out particular models and rejecting alternatives. Most abhorrent to him were alternate solutions that contained singularities: points in which physical quantities such as density blow up to infinity and space-time simply seems to cease, like the termini of dead-end streets. From such anomalies, anything could emerge. In his pursuit of universal order, Einstein detested such loose ends.

In 1965, ten years after Einstein's death, Cambridge physicist Stephen Hawking theoretically determined that all physically reasonable cosmological models must begin in a singularity. Given that any type of model, no matter how irregular, might pop out of a singularity, Hawking's monumental discovery suddenly brought all of the anisotropic (expanding differently in various directions) and high-curvature alternate solutions of general relativity into sharp focus. At the same time, increasingly detailed astronomical observations of the universe confirmed its near-isotropy (sameness in all sky directions) and close-to-flatness (zero curvature) on its largest scale. The mismatch between theory and observation was glaring. In short order, finding ways to start with sheer mayhem and end up with stodgy regularity became a top priority for cosmologists.

Two of the innovative physicists leading the charge had close connections with John Wheeler. In the late 1960s, Charles Misner, his former student, focused on a dynamic (how forces and energy fields affect motion) approach to smoothing out the matter and radiation in the universe. Because of his model's mixing properties, he dubbed his proposed solution the "Mixmaster universe" after a popular food processor. Shortly thereafter, Brandon Carter, who, though based at Cambridge, spent ample time as a research visitor at Princeton under Wheeler's sponsorship and remained a regular correspondent, coined the "Anthropic Principle" as a selection mechanism for distinguishing cosmological models and conditions based on their propensity for supporting the emergence of intelligent life able to observe the cosmos. By hook or by crook—dynamics or a selection principle—cosmologists hoped to fulfill Einstein's goal of a comprehensive explanation as to how general relativity offered a blueprint for the universe's conditions.

SKETCHES OF THE COSMOS

Einstein's disappointment that general relativity failed to produce a unique solution representing the observed cosmos went back to his earliest attempts to apply his theory to cosmology. In 1917, he proposed his first model of the universe at large, hoping it would offer a self-consistent explanation of the stability of space. At the time, he thought wrongly that the universe, as a whole, must be unchanging over time. Initially, though, he could find only solutions that would either expand or contract. Hastily, he added the cosmological constant, an ad hoc term counteracting gravity, to the equations of general relativity for the purpose of ensuring a static result. Publishing his stable solution that year, he thought wrongly that it would be the end of the story.

Much to Einstein's consternation, however, later in 1917 Dutch physicist Willem de Sitter found an alternative, matter-free cosmology that behaved in a markedly different manner. Instead of keeping still, space did the exact opposite. Driven incessantly by the repulsive power of the cosmological constant, while lacking gravitational clumping for counterbalance, it expanded exponentially. De Sitter's finding opened the Pandora's box of offbeat solutions to general relativity.

More than a decade later, Edwin Hubble's 1929 telescopic evidence of the recession of distant galaxies away from ours convinced Einstein and many others that the cosmos is expanding and therefore the stabilizing term was unnecessary. Einstein decided to remove the cosmological constant from general relativity, deeming its initial inclusion a blunder. More recently, that term has been revived as a way of modeling the accelerating expansion of the universe. At any rate, with or without the term, the range of general relativistic solutions is enormous.

Mathematical options are not necessarily physically realistic, however. Therefore, automatically tagging that assortment of cosmological models as a multiverse of parallel realities would be misleading, unless one takes the expansive view that everything mathematically conceivable is also physically possible. Einstein himself came to realize that there were many untenable solutions to general relativity. He summarily dismissed those that contained oddities, or didn't match his expectations for other reasons, as simply being mathematical artifacts. He didn't perceive them as existing in any sense, let alone being part of a multiverse.

Rather, in studying the universe, Einstein and his contemporaries focused almost exclusively on the small subset of isotropic, homogeneous models among the hodgepodge of possibilities. By assuming isotropy and homogeneity (uniformity from point to point in space), the number of cosmological solutions narrows down to a handful. That was fine with Einstein, who was mainly interested in showing that general relativity had viable solutions matching the observed universe. At first he wished to model the traditional, static cosmos, and later, the spatial growth suggested by Hubble's data.

In 1932, after Hubble's evidence for universal expansion rocked the world of astronomy, Einstein joined with de Sitter to craft a simple way of representing a ballooning cosmos. They proposed an isotropic, homogeneous, flat model without a cosmological constant. It would grow forever, but continuously slow down in its expansion rate. The Einstein-de Sitter model, along with similar predecessor models developed by Russian physicist Alexander Friedmann and Belgian mathematician and priest Georges Lemaître, profoundly influenced the direction of cosmology, and offered an elegant spatial framework for what later became known as the Big Bang theory. (Ironically, the name was coined in 1949 by one of the theory's principal detractors, Fred Hoyle.)

Once those straightforward isotropic, homogeneous Big Bang models were developed, it seemed only a matter of time before observational cosmology would match the characteristics of one of them, and the game essentially would be over. Looking at the full range of general relativistic solutions, including messy ones, would merely be a mathematical exercise, many physicists thought. A review article published in 1933 by Princeton physicist Howard P. Robertson set the tone, by focusing on the properties of the range of spatially uniform solutions. Because of Robertson's influential cataloguing of homogeneous, isotropic cosmologies, supplemented by work by the British physicist Arthur Walker, along with the earlier research of Friedmann and Lemaître, the various growth patterns over time of such spatially uniform models are called Friedmann-Lemaître-Robertson-Walker (FLRW) metrics.

Tracing the expansion of space backward over the eons, following any variant of the FLRW metric, everything we see in the observable universe was once incredibly compact, and hence immensely hot.

Moving forward in time from that fiery beginning leads to an unmistakable conclusion. If the universe expanded from a super-hot, ultra-dense fireball, then the cooled-down, leftover radiation should be all around us. There would be nowhere else for it to go. That turned out to be the Cosmic Microwave Background Radiation (CMBR) of about 2.73 degrees Kelvin (above absolute zero). In 1964, Arno Penzias and Robert Wilson, using the Holmdel horn antenna, first detected the CMBR as a steady, omnipresent hiss in all directions, no matter which way they aimed their instrument. Bob Dicke and P. James E. "Jim" Peebles would successfully interpret its cosmic significance the following year.

None of those theoretical and experimental developments would seem to inspire anything like a multiverse. Rather, the predominant conception, at that point, was that of a single isotropic, homogeneous FLRW cosmology that started off hot and dense, and left its telltale imprint in the CMBR. If anything, that radiation profile was too smooth. Researchers had yet to identify the minor variations of long ago, recorded in the CMBR, that would eventually form the seeds of structures such as stars and galaxies. Otherwise, the story of a unique universe born in a Big Bang seemed convincing.

Yet, the results did not quite satisfy a group of physicists who, in the spirit of earlier work by Arthur Eddington and Paul Dirac, were obsessed by the idea of cosmic coincidences among the constants of nature. Eddington, noted for helping organize the eclipse expeditions that supported general-relativistic light-bending predictions, and Dirac, a respected Nobel laureate, each had tried to find deep, fundamental reasons for the combinations of values of certain physical constants. Ironically, for someone known for helping verify the Big Bang, Dicke followed closely in their footsteps. Though trained in experimental radio wave physics, he often speculated about the special conditions of the universe—including how parameters such as G, the universal gravitational constant, c, the speed of light, t, the age of the universe, and others relate to each other. He postulated that there was a link between those combinations of constants and the emergence of humanity. If those parameters were very different, structures that ultimately led to living planets such as Earth might not have had the opportunity to develop. For instance, if gravity was stronger and the universe simply re-collapsed or if it was weaker and the universe expanded too quickly, the processes of stellar and planetary formation might never have transpired.

Such ruminations about why the universe possesses particular physical parameters conjure up pictures of alternative scenarios—whether real or imagined. Thus, it potentially—but not necessarily—opens the gate to multiverse ideas through speculations about a grand cosmic competition. Before making such a leap, though, prudence would seem to demand justifying such values based on dynamics within a lone universe.

Dicke's attempt to do the latter involved modifying standard general relativity. He subscribed to the notion that G should be replaced by a scalar field: an entity that varies from era to era and location to location. Such cosmologies would become known as "Brans-Dicke," named for Dicke and his collaborator (and student of Misner's) Carl Brans, or sometimes "Jordan-Brans-Dicke," to recognize German physicist Pascual Jordan's earlier, independent embrace of a similar concept of changing G.

Along with his notion that gravitational strength has altered throughout the ages, Dicke believed in a cosmic history encompassing multiple cycles of creation and destruction. Rather than a single Big Bang, the burst of matter and energy was part of an oscillating pattern of contraction and expansion. As physicist Richard Tolman had shown, such models result in a buildup of disorderly energy from era to era. Dicke thought, therefore, that the CMBR was the cooled-down remnant of waste energy from a previous cosmic cycle.

Belief in oscillatory models of the universe was rather common from the 1930s up until the early 1960s. That's because it offered a reasonable answer to the question of how the matter and energy of the Big Bang emerged. Either that material had been there forever, and suddenly started to grow, or else the expansion was the outcome of a prior contraction. Oscillatory models nicely echoed the cyclic views expressed in the Vedic writings of Hinduism and many other notions in Eastern philosophy. Unlike Nietzsche's eternal return, each cycle would be different from the previous one. Soothingly, there would never be a true beginning or a true ending; rather, cosmic history would constitute an endless procession of universal reincarnation. Dicke was well aware that his oscillatory ideas matched Eastern views, and found those common themes a comforting way of looking at cosmic creation and destruction. Much to his disappointment, however, "singularity theorems" newly derived in general relativity soon began to call into question the notion that anything could have preceded the Big Bang.

BORN THIS WAY: THE SINGULARITY OF CREATION

If light rays are the highways of information exchange in the universe, singularities comprise abrupt dead ends. Such bizarre points or regions in which spatial curvature is infinite, spatial and/or time coordinates are undefined, and density and other physical quantities blow up to infinity offer the ultimate scientific enigmas. No wonder Einstein, in attempting various types of unified theories, avoided them like the plague.

In 1916, physicist Karl Schwarzschild, while serving in the German army during World War I, developed one of the first general relativistic solutions—representing how gravity acts in the vicinity of a static ball of matter, such as an astral body. That solution behaves very strangely for ultra-high densities—a state reached by the catastrophically collapsed cores of extremely massive stars. Starting in 1967, Wheeler began calling them "black holes," suggested to him by an audience member at one of his talks, and thereby popularized that expression.

For such objects, an invisible frontier, called the event horizon, emerges at a particular radius. In some, but not all, coordinate systems, their space and time values are off the charts—rendering them removable singularities. Removable singularities constitute mathematical artifacts rather than physical dead ends.

According to classical relativistic theory, once inside the spherical region bordered by the event horizon, nothing might escape, not even light. Light and materials would pass through just fine in the inward direction, but not outward. If one applies quantum rules to gravitation, though, there is a way—a process called Hawking radiation—for energy gradually to be emitted into space. Therefore, in reality, the event horizon is a semi-permeable membrane, rather than a zone of no return.

Black holes' dead centers, in contrast, are inescapable and absolutely lethal. They possess infinite curvature and infinite density, and offer crushing finality to anything doomed to approach them. Such physical singularities aren't removable, no matter which coordinate system is used.

On the other hand, in some coordinate systems black hole solutions mathematically extend into another spatial sheet, curiously enough, like the two halves of an hourglass. Together they represent a kind of portal connecting one part of space to another; albeit a wholly impassible one, as if two rooms in a building were joined by a lethal trap door that no one dare attempt to traverse. One might think of the original black hole solution as a kind of entranceway, or "mouth," the continuation as a second "mouth," and the connection between the two as a mind-bogglingly narrow "throat."

In some theoretical conceptions, particularly a model advocated by physicist Lee Smolin, that conjoined second space serves as the diametric opposite of catastrophic stellar collapse. Rather, it offers explosive expansion—the birth of a new baby universe. Hence, in that view, any universe with stars massive enough that their cores suddenly implode into black holes effectively serves as a type of multiverse. Rather than parallel branches of reality, however, such a multiverse would constitute a nursery for baby universes. Some of those little ones, if they possessed the right blend of physical parameters, would grow up into grand universes with massive stars themselves, birthing even more baby universes. Smolin pointed out that the situation would resemble survival of the fittest in an evolutionary battle between universes, with reproductive success based on the number of stars that complete the full path to collapse. Those stars would need to be in a universe with the right blend of physical constants, helping explain their auspicious values. Note, however, that while the existence of black holes has long been confirmed, the notion of those bodies spawning baby universes in a Darwinian competition remains extremely hypothetical—a minority view even among multiverse advocates.

Schwarzschild black holes don't spin and are electrically neutral. However, they aren't the only type of black hole. Over the years, researchers have discovered other theoretical solutions to general relativity that model different black hole varieties. Months after Schwarzschild published his model, German engineer Hans Reissner discovered a solution taking into account electric charge. Two years later, Gunnar Nordström independently derived an equivalent solution, now known as Reissner-Nordström black holes. Because massive celestial objects are generally electrically neutral, such charged models have more of a theoretical significance than a connection with what is out there.

More significantly, in 1963 New Zealand–born astrophysicist Roy Kerr presented at a Texas conference the first rotating black hole solution. It represented twirling, electrically neutral, ultradense massive objects. Instead of a point singularity at their centers, such spinning maelstroms contain ring singularities—lethal loops of infinite curvature and density. The astrophysics community immediately realized the profound importance of such rotating solutions. Most, if not all, black holes should rotate. Black holes' less massive cousins, neutron stars, certainly do—as discovered in 1967 by Northern Irish astrophysicist Jocelyn Bell Burnell in her identification of rapidly spinning, astral radio beacons called "pulsars."

According to the law of conservation of angular momentum, if something is revolving and shrinking, it should spin faster and faster as it becomes increasingly compact. That's what happens in the death of ultra-massive stars. After their primary sources of nuclear fuel are expended, their cores suddenly collapse and increase their rate of spin. If those cores are sufficiently massive, gravitation overcomes all resistance, and they catastrophically crunch down into black holes. Their residual angular momentum would thereby make them whirl into a frenzied vortex. Rotating Kerr solutions, in contrast to static Schwarzschild solutions (or charged, stationary Reissner-Nordström models), match that gyration expectation.

Based on Kerr's work, in 1965 Ezra "Ted" Newman, an American physicist based at the University of Pittsburgh, threw every major ingredient into the pot and tackled the question of charged, spinning black holes. Based on their mass, charge, and angular momentum, and an assumption of spherical symmetry, he found exact black hole solutions with either ring or point singularities, generally (but not always) surrounded by event horizons. From that point on, researchers would consider the other main theoretical varieties of black holes (at least those constituting collapsed stellar cores) to be subcategories of the comprehensive Kerr-Newman solution, as it is called.

In those discussions of black holes from the early-to-mid 1960s, a critical question emerged: How do we know that the cores of physical stars—with the imperfections present in all natural bodies—collapse precisely into perfect spheres and thus represent variations of the Kerr-Newman solution? Might they end up irregular enough to avoid having singularities altogether? After all, when familiar objects implode the outcome is often messy, not a perfect sphere. Imagine poking a pin into a balloon or basketball, and watching them deflate. In either case, once all the air is released, some kind of misshapen glob would emerge, rather than a smooth ball.

Remarkably—in work that would be honored decades later with the 2020 Nobel Prize in Physics—in 1965 British mathematician and mathematical physicist Roger Penrose developed powerful techniques in general relativity to analyze the properties of "closed trapped surfaces" (regions that prohibit communication, via light or other signals, to other parts of space-time), such as black hole vicinities. Making use of mathematical theorems developed by Indian physicist Amal Kumar Raychaudhuri, he proved that singularities lie at the heart of all black holes. The reason we don't see them directly in nature is that they're cloaked behind event horizons and other types of closed trapped surfaces.

Several months later, young Cambridge-based physicist Stephen Hawking had an extraordinary insight based on Penrose's results. Through the magic of time-reversal, he imagined running the Big Bang backward into the distant past and examining its nascent moments. Curiously—with its dynamics in rewind mode—its matter and energy would unite into an ultradense state in a manner somewhat similar to black hole collapse. Applying Penrose's work to the time-reversed Big Bang, Hawking found that Einstein's equations of general relativity mandate that the cosmos began in a singularity. From that point of infinite density everything in existence was hatched.

Hawking's PhD supervisor, Dennis Sciama, soon informed

Image 13. British theoretician Stephen Hawking, who developed many key theorems in general relativity, astrophysics, and cosmology. Credit: The Franklin Institute, courtesy of AIP Emilio Segrè Visual Archives, Physics Today Collection.

Dicke about his discovery. Dicke realized that it challenged his notion of energy remaining from prior cosmic cycles. Nothing could persist through a cosmological space-time singularity—given that all roads of communication begin there (similar to all paths ending in the crushing presence of black hole singularities). Unless there was a way around that quandary, for all practical purposes, the Big Bang would constitute the very beginning of the cosmic timeline. Anything before it—in the event that there were prior cycles—would lack a causal link to our universe. Therefore, Dicke's conjecture that the CMBR signaled a prior incarnation of the universe lost credibility in favor of explanations involving a hot primordial fireball that somehow emerged from an initial singularity, expanded, and cooled down in the process, until constituting the cold, present-day radio hiss that fills all of space.

LOST HORIZON

In tracing back cosmic history, astrophysicists soon realized that for hundreds of thousands of years after the Big Bang the universe must have been impervious to light. That's because it was too hot in that primeval epoch to allow for neutral atoms. As George Gamow, Ralph Alpher, and Robert Herman had shown in the late 1940s, and Peebles had calculated with more realistic parameters in the mid-1960s, the fiery first three minutes of the universe produced a soup of photons, electrons, positive ions, and other particles. Such ions included the nuclei of simple hydrogen (single protons), deuterium (one proton bound by the strong force with one neutron), tritium (one proton bound by the strong force with two neutrons), helium-three (two protons bound by the strong force with one neutron), helium-four (two protons bound by the strong force with two neutrons), and a small quantity of lithium. The photons ricocheted between the charged particles with no chance of freedom, causing the cosmos to be opaque.

Then, around 380 thousand years after the Big Bang in the

recombination era, the universe finally cooled down enough for neutral atoms to form. Ions greedily grabbed up electrons and allowed themselves to calm down and become more stable—rather than constantly exchanging energy in a colossal pinball game. Consequently, photons became free to travel long distances. Space, from that point on, was transparent to light. Then, over billions of years, as space steadily expanded, those photons shifted to lower frequencies corresponding to a continuous drop in temperature. Ultimately, they became the frigid constituents of the CMBR, and showed up in Penzias and Wilson's horn antenna data, along with numerous other radio detector results since then.

Meanwhile, the matter abandoned by the sea of radiation took on a life of its own. Thanks in part to an unknown, invisible agent called "dark matter," and the clustering power of gravitation, denser regions began to gather into larger and larger chunks. Another unidentified ingredient called "dark energy," sometimes represented by including the cosmological constant term in general relativity, opposed gravitation, and threatened to accelerate spatial expansion. However, because of its low value, astrophysicists believed that it played little to no part during the nascent era of the universe, readily allowing gravity to clump material without impediment. Because of that coagulation, we know that some parts of the primordial sky must have been at least slightly more concentrated than other regions, leading gravity to concentrate them even further.

At some point, bodies began to form that were massive enough and hot enough in their cores to fuse hydrogen into deuterium and other light isotopes and elements. The first stars—gargantuan bodies—ignited, and the universe began to light up in a different way. Within the cooling, almost-uniform glow, shifting over time into lower and lower frequencies, were distinct, shining stars enacting cycles of fusion that transform hydrogen in a multistage process into helium. Ultimately that generation of stars exploded, creating heavier elements in its wake, and seeding a second and third family of stars—though mainly made of hydrogen, some helium, and their respective isotopes, each now had a larger percentage of "metals" (what astronomers call elements beyond hydrogen and helium) than those first stars.

In their detective work unraveling the multi-billion-year story just described, astronomers and astrophysicists have relied on a variety of celestial clues. Radiometers, dating back to the Penzias-Wilson apparatus, followed by a device built by Dicke's Princeton group, and later by balloons and satellites, have mapped out the CMBR with increasing precision. Meanwhile, terrestrial and space-based telescopes—from the Hooker telescope at the Mount Wilson Observatory, used by Hubble and others in the 1920s, to the James Webb Space Telescope launched in late 2021—have captured the stars, planets, gas clouds, galaxies, clusters of galaxies, and other astral bodies and structures in space.

By matching the radiative and material profiles, key questions have arisen. When Penzias and Wilson revealed the extraordinary temperature uniformity of the CMBR, such sameness, at first glance, boded well for assigning it a cosmological, rather than local, origin. After all, if it was the result of street noise, stellar activity, or a galactic burst, all of those things would be found in a particular direction, rather than being the same at all angles. However, once astrophysicists began to contemplate the CMBR's origin in the recombination era, its absolute uniformity became suspect. Small density fluctuations in matter would have been needed at that time to form the seeds of what later would become stars and other formations. Corresponding to those irregularities, there would need to be minute temperature variations in the CMBR, associated with the concentrating effects of matter on radiation in its vicinity, making it slightly hotter. Fortunately, subsequent probes of the CMBR, such as the COBE (Cosmic Background Explorer), WMAP (Wilkinson Microwave Anisotropy Probe), and Planck satellites, revealed those minute temperature fluctuations—a few parts in one hundred thousand—in increasingly precise detail. Those images of hotter and colder spots have come to be known as the "baby pictures" of the universe.

Aside from the need for subtle temperature variations, however, an even trickier conundrum emerged in the late 1960s, largely thanks to the insights of Charles Misner. Dubbed the "horizon problem," it had to do with tracing the photons in all sky directions—as seen in maps of the CMBR—backward in time to their origin in the recombination era.

Normally, temperature uniformity is a sign of thermal contact. If you see a dinner table set with steaming bowls of soup, all seeming to be around the same temperature, your natural assumption is that they were all just ladled out of the same pot. There, the batch of soup, once heated, would have had time and opportunity to rise to an even temperature while remaining in thermal contact. On the other hand, if on the same dinner table one of the bowls of soup is ice-cold, you might assume that it wasn't just taken from the same pot. Rather, perhaps it was mistakenly poured straight out of a container that had been in the refrigerator without having been heated up. In other words, it wasn't in thermal contact with the heated pot of soup.

In the late 1960s, by imagining running cosmic expansion in reverse, Misner traced the CMBR profiles of various sky regions back to the time of the recombination era when those photons were first released into space following the creation of neutral atoms. Surprisingly, based on FLRW models expanding at the range of rates established by Hubble's law (the

Image 14. American physicist Charles Misner, who made major contributions to gravitational theory from the 1950s until the present day. Credit: AIP Emilio Segrè Visual Archives, Physics Today Collection.

farther away a remote galaxy, the faster it is moving away from us), he calculated that the release locations corresponding to points in the sky currently separated by greater than a thirty-degree angle would represent places that were never in thermal contact. In our analogy that would be like soup from the heated pot versus soup from the refrigerated container. In that case, how could the cosmic radiation be so astonishingly uniform in temperature on average—despite minute variations of a few parts in one hundred thousand—in the present day? Coincidences of that magnitude don't typically happen in science.

The term "horizon problem" has to do with communication limits. Light, with all of its swiftness, nevertheless takes a certain amount of time to travel. Therefore, during any given time interval, it has a maximum reach. If, during the development of the universe, light from one point in space could not possibly have had enough time to interact with that of another point in space, astronomers say that they are beyond each other's horizons. Misner calculated that at no time in the past history of the universe would the light from one sky direction have had the opportunity to circumnavigate space and reach points in other sky directions more than thirty degrees away. Therefore, the fact that the temperatures around the celestial dome are so close to each other cannot be explained by means of simple Big Bang cosmologies, based on FLRW metrics, alone. Such dynamics do not allow sufficient opportunity for distant sky points to have even been close enough for thermal exchange.

To return to our soup analogy, imagine a defective stove with burners heating at different rates—some searing with raging flames, others slow-cooking like dying campfire embers. If a soupy concoction is heated (or not) in separate pots on the various burners, and ladled out directly to different bowls, their temperatures would be most uneven. However, if the contents of each pot were poured into a mixer or food processor—such as a vintage Sunbeam Mixmaster blender—in seconds all the contents would have an even temperature and consistency. If left to simmer for a while (say in a large common pot) and then doled out into bowls, everyone could savor uniformly heated fare. Could the cosmos similarly have had a mixing era, wondered Misner, in which its temperature evened out?

Misner realized that seeking a solution to the horizon problem by envisioning very different cosmic dynamics in the past would be a novel way of representing the universe. As he later noted, cosmologists used to ask, "Can we make a model of the universe that is consistent with observation?"

"A better challenge for cosmology," Misner continued, "is that there is some way that the universe would sort itself out into the kind of universe we see."[2]

Though dear friends and fellow former PhD students of Wheeler, clearly Everett and Misner maintained radically different approaches toward resolving conundrums in fundamental physics. While Everett was very comfortable with a multiverse of near-replicating realities, Misner advocated theories constrained to a single universe, albeit possibly changing in dynamics over time.

Wheeler clearly ended up on Misner's side—in terms of one versus many—ultimately arguing that the universe, though beginning in mayhem, became uniquely suited to humanity via a self-consistent loop involving human measurement of the past. However, he liked to be polite to and supportive of his former students. When chatting with Everett and Misner a few weeks after Everett's 1977 talk at the University of Texas, Wheeler professed to be only mildly conflicted about Everett's theory, stating that he "mostly believed his interpretation but reserved Tuesdays once a month to disbelieve it."[3] That was clearly a way of softening his skepticism as not to offend.

In general, the difference between those who embrace or avoid multiverse schemes is in some ways akin to athletes driven by competition (even against unseen adversaries, such as in a virtual simulation) versus fitness enthusiasts who motivate themselves to change without even thinking about others. After all, just as a prisoner might build himself up while in solitary confinement, conceivably a lone universe might singularly evolve from flabby irregularities to a toned, uniform state if the dynamics are just right. If such a sole metamorphosis is possible, why add the complications of comparisons?

CHAOS IN THE COSMIC CAULDRON

Misner's identification of a cosmological model with mixing properties that would potentially smooth out the temperature of the early universe and thereby resolve the horizon problem derived from his broad familiarity with anisotropic solutions in general relativity. His "boot camp" for that effort was his extensive experience working with Wheeler in characterizing the wide range of geometries that solved Einstein's general-relativistic equations under various physical circumstances—from vacuum (as close to emptiness as possible) conditions to various distributions of matter and energy. His explorations aligned with Wheeler's quest for a quantum theory of gravitation based on a deep understanding of the role of geometry in the cosmos. In Wheeler's view, just as positions and momenta are each generally smeared in the quantum state representations of elementary particles, different geometric configurations would be blended together in a quantum theory of gravitation.

The speculative scenario Wheeler pursued, and to which he had introduced Misner, was to imagine the primitive cosmos as a jumble of competing geometries—a space-time foam, or quantum foam, in his parlance. That drama would take place primarily at the Planck scale—that is, in regions the size of the Planck length, approximately 10^{-35} meters. Within such a minuscule, undetectable domain, quantum gravity is thought to prevail. In such a turbulent era, reality would be a quantum superposition of myriad kinds of spatial configurations.

Wheeler dubbed the abstract space of all possible three-dimensional geometries "superspace."[4] Such wouldn't necessarily be a multiverse, but rather could be the playing ground for a "sum over histories" in Hilbert space, in the spirit of Feynman's work in quantum electrodynamics.

Along with his Planck-scale ruminations, Wheeler explored nonconventional arrangements in general relativity—denizens of superspace—sometimes in conjunction with Misner, Dieter Brill, American gravitational physicist Kip Thorne, and other students in his gravitational research group. His ambitious goal was to replace the world of elementary particles with purely geometric structures. Some of the oddities he considered were "geons"—energy fields (such as electromagnetic fields) bound together by their own self-gravitation—and "wormholes," spatial shortcuts through which such fields could be threaded, like using a needle to push twine through fabric. The name of the latter derives from the tunnels made by worms through various materials, such as apples and old books. Its theoretical origins hark back to a 1935 proposal by Einstein and Nathan Rosen for "bridges" connecting different segments of space-time—based on extensions of Schwarzschild solutions into another domain, similar to the two halves of an hourglass.

Although the bizarre solutions of general relativity explored by Wheeler were astronomical in size, he hoped for a theoretical way to somehow shrink them down enormously—to particle dimensions, or maybe even further to the Planck scale, where they'd be constituents of quantum foam. He made his case so ardently at the 1957 Chapel Hill meeting that Feynman jokingly nicknamed him "Geon Wheeler."[5]

Under Wheeler's guidance at Princeton, and afterward on his own, as a professor at the University of Maryland starting in 1963, Misner made great strides in broadening scientific understanding of classical general relativity and its potpourri of offbeat solutions. By delving into wormholes and other unconventional models, as part of excursions into superspace, Misner became comfortable with wielding Einstein's mathematical machinery in novel ways, far beyond the standard FLRW Big Bang approach. Interactions with mathematician Abraham Taub, who had studied under Robertson's guidance in the early 1930s, also inspired him. Taub had developed one of the first anisotropic (behaving distinctly in different directions) cosmological models, slightly more complex than an earlier anisotropic model developed by American mathematician Edward Kasner in 1921. Kasner is perhaps best known for popularizing the term "googol" for the numeral 1 followed by 100 zeroes—a term coined by his nine-year-old nephew.

As Misner came to learn, homogeneous, anisotropic geometries were first classified by Italian mathematician Luigi Bianchi in 1898 according to a mathematical field known as group theory. Group theory examines the workings of operations on a set. For example, adding two integers does not depend on the order of operations, but three-dimensional rotations do. Therefore, addition of integers and three-dimensional rotations constitute two distinct groups. Bianchi's classification ranges from the simple "Type I" to the intricate "Type IX," each representing increasingly complex relationships in group theory.

Kasner's vacuum (matter-less) solution, based on Type I, is essentially the Big Bang with directional bias. Instead of expanding at identical rates in all three spatial dimensions, it grows in two directions, generally at different rates, and shrinks in the third direction. The rates of growth and shrinkage are tied together by simple relationships. Taub's vacuum model, based on Type II, is somewhat more complicated in its dynamics.

In the late 1960s, Misner, and independently a team of researchers based at the Landau Institute for Theoretical Physics in Russia—Vladimir Belinsky, Isaak Khalatnikov, and (later) Evgeny Lifshitz, known as "BKL"—decided to focus on an anisotropic vacuum cosmology, based on Bianchi Type IX, with intricate dynamics resembling a blender. Christening it the "Mixmaster universe," Misner saw it as a promising candidate for resolving the horizon problem by positing an era of mixing early on in cosmic history. He used Hamiltonian (energy conservation) principles to model its churning dynamics, with the goal of smoothing out the temperatures of the primordial cosmos and justifying today's near-uniformity.

Misner recalled Wheeler's reaction to the Mixmaster universe idea as very positive. As he remembered, "Wheeler was very excited about it and very happy about it."[6]

The Mixmaster universe was a valiant effort, but ultimately it didn't mix up space quite enough to permit temperatures to equalize. Unlike the evenly blended soup in our analogy, segments of the CMBR would remain much hotter, and other parts much colder than what is physically observed. Therefore, unfortunately, it left the horizon problem unsolved.

The BKL team took a somewhat different approach, centered on examining how the dynamics of a vacuum Type IX model would behave if run backward in time toward the initial singularity of the universe. By making certain approximations, they demonstrated that its time-reversed behavior would pass through a series of "Kasner eras" in which two of its directions would oscillate (take turns growing and shrinking, like a pair of jack-in-the-box toys popping up and down in contrary ways), while the third uniformly contracted. In other words, it would behave like a succession of Kasner Type I models. After a number of those oscillations, called an "epoch," the overall directions would switch. The formerly contracting scale factor would change places with one of the oscillating ones, like a sole dancer cutting in to borrow the partner of another to practice a different kind of move. Such patterns of oscillations, and directional shifts, would happen again and again in the approach to the initial singularity.

Strangely enough, as pointed out and examined by British physicist John Barrow in a 1982 issue of the respected journal *Physics Reports*, if one keeps a running count of the number of Kasner eras for each epoch, the figures vary widely and seemingly randomly. While, as a solution of general relativity, it is emphatically a deterministic model, it seems to behave as haphazardly as rolling a die or flipping a coin (of course, those on a fundamental, classical level are deterministic, too). In the "chaotic cosmology programme," Barrow wondered if the early universe could have been in a chaotic jumble, and somehow smoothed out over time. Alternatively, he pondered scenarios (in some ways akin to Wheeler's quantum foam proposal) in which a primordial superspace housed an array of universes of various tangled geometries, and somehow we ended up in a nearly homogeneous, isotropic sector—perhaps because such conditions are necessary for the emergence of life.

The contemporary idea of deterministic chaos dates back to the discovery by meteorologist Edward Lorenz, published in a 1963 paper, about the sensitivity of weather forecasts to initial conditions. In a phenomenon he called the "butterfly effect," he noted that even the flapping of a butterfly's wings in one part of the world could potentially cascade into a major effect in another part. His discovery stemmed from developing a computer model of the weather—based on temperatures, pressures, wind velocities, and such in various locations—inadvertently entering in very slightly different numerical data, running his program, and witnessing it output a profoundly different forecast. Other physicists explored similar sensitivities to initial parameters in a diverse range of deterministic models. In 1975, mathematician James Yorke of the University of Maryland introduced the term "chaos" to characterize such deterministic models exhibiting seemingly random behavior, a fitting appellation that stuck.

Chaotic cosmologies reveal the astonishing diversity of the solutions to Einstein's gravitational theory. Given the limitations of current probes in mapping out the observable universe, and the likelihood of vast regions of space beyond observability, the jury is still out on whether or not anisotropic cosmologies and other unconventional solutions beyond the FLRW have physical meaning, or alternatively serve simply as fascinating mathematical models.

FLAT-OUT TRUTH

Numerology would seem to have no place in physics. Yet, on occasion, numerical patterns have led to important discoveries. For example, in the early 1960s, physicist Murray Gell-Mann looked at the properties of certain elementary particles, assigned them into arrays on that basis, and deduced the existence of other, hitherto unknown, particles—which were later discovered.

When Nobel laureate Paul Dirac, one of the great theoretical physics geniuses who successfully predicted antimatter (particle counterparts of opposite electric charge, such as positrons being the positively charged versions of electrons), announced the "Large Numbers Hypothesis" (LNH) in 1937, a segment of theorists took him very seriously. Based partly on earlier speculation by Arthur Eddington, Dirac's proposal offers a way of understanding key features in the universe by connecting certain extremely large ratios—each of the order 10^{40} (1 followed by 40 zeroes)—of combinations of physical constants. He surmised that what we think of as constants could change over time, as long as those large ratios are preserved. For example, the ratio between the age of the universe and the time it takes for light to cross the classical radius of an electron is comparable to the ratio of the electric and gravitational attractions of an electron to a proton, each in the ballpark of 10^{40}. Square that humongous number, and one obtains 10^{80}, approximately the number of nucleons (protons and neutrons) in the observable universe. Dirac conjectured that to maintain those coincidences, gravitational strength must weaken over time.

By the late 1950s and early 1960s, Dirac's idea had found an important exponent in Dicke, who was investigating alternatives to general relativity. Dicke's embrace of the notion that G, the gravitational constant, should be replaced by a varying scalar field—that is, Brans-Dicke theory—was a direct consequence of his interest in Dirac's LNH. Unlike Dirac, Dicke was an experimentalist, and sought to test his gravitational hypotheses by means of physical and astronomical observations. As Peebles recalled, Dicke once made a friendly bet with Wheeler about gravitational light-bending results, with the former hoping it would reveal cracks in general relativity's predictions and the latter expecting that general relativity would be vindicated.[7] Unfortunately for Dicke, general relativity has triumphed, so far, in every single test. There is no evidence for a changing gravitational constant.

Regardless of the validity of his hypothesis, Dicke contributed valuable insight to cosmology by emphasizing how the mere fact that we are here—and life has flourished on Earth in general—places strict bounds on the ways the universe could have developed. His analysis was an important forerunner of what became known as the Anthropic Principle. Take, for example, the LNH. Dicke believed such enormous figures reflected the stage of the universe we find ourselves in, one involving far-more-complex evolutionary processes than the rudimentary activity of elementary particles on the atomic scale. The many steps required for evolution sets a

Image 15. Experimental radio wave and gravitational physicist Robert H. Dicke, who helped interpret the first evidence of cosmic microwave background radiation and who advocated for a form of the Anthropic Principle. Credit: AIP Emilio Segrè Visual Archives.

time scale for the emergence of humanity much greater than that of atomic transitions, leading to a very large number as their ratio. As he noted in a 1957 paper, "Man with all his complexity could not have evolved in a characteristic atomic time."[8]

Sometime in the early 1960s, when Misner was still at Princeton, Dicke discussed with him another cosmic situation that bears on the existence of life. In what Misner dubbed the "Dicke paradox," Dicke pointed out the favorable situation for living beings that the geometry of the universe is very close to flat. In isotropic, FLRW models, there are three geometric options: one with positive curvature, one with negative curvature, and one with zero curvature, also known as "flat." Anisotropic and inhomogeneous models have many more options for curvature—and can be as wrinkled as dormitory sheets crumpled in a heap on the floor. Despite all of those possibilities, Dicke noted that the universe today seems to be very close to flat. That's a good thing, because extreme positive curvature would have led to the universe collapsing well before there was time for stars with planets like Earth to develop. Extreme negative curvature would have produced much faster early cosmic expansion, preventing gravitational clumping into stars and galaxies. The fact that we're here matches well with a close-to-flat cosmos.

Misner's explorations of anisotropic models, and the discovery of chaotic behavior in some of them (not only Bianchi Type IX, but also Type VIII, displays chaos in vacuum situations) invited the question of what range of early conditions could have led to the present-day close-to-isotropic, nearly flat situation. After all, the initial singularity could have spewed out a universe of any possible geometry. One might imagine a multiverse of competing geometries, some starting out as flat as a pancake, others as crinkled as a shar-pei's skin, each vying to evolve into a universe flat and even enough to resemble our own. Which of the wrinkly ones would succeed in flattening out?

The answer, as pointed out in work by mathematician Christopher Barry Collins and Stephen Hawking, as well as by Dicke and Peebles, is most surprising. In a 1973 paper, "Why Is the Universe Isotropic?" Collins and Hawking examined the full gamut of anisotropic models and found that they tend to get less isotropic over time. Their irregularities grow, rather than shrinking, leading to increasing unevenness. In fact, the chance of an anisotropic cosmic beginning, such as the Mixmaster universe, smoothing out over the eons is effectively nil. Only a completely isotropic start, it would seem, would lead to an isotropic outcome, they argued, unless there is a selection mechanism singling out such models.

Later, Dicke and Peebles calculated that to produce the current near-flatness, gauged by a parameter called "omega" that measures flatness of our universe at a given time (with one meaning perfectly flat), at one second after the Big Bang that same parameter would have had to be equal to one within one part in 10^{15} (1 followed by 15 zeroes).[9] In other words, if the primordial universe was even mildly curved, by now it would have become extremely curved rather than reach the state of near-flatness we observe. Their striking results drove home the point that flatness truly needs to be fine-tuned. The dilemma they posed became known as the "flatness problem."

Referring to Dicke, and Brandon Carter, whom they knew at Cambridge, Collins and Hawking concluded their paper with the suggestion that there are multiple universes (embedded in what later would be called a "multiverse"), among which at least one—ours—has been isotropic and flat enough to allow time for stars like the sun to form, orbited by planets housing intelligent life. If we didn't live in such a life-friendly universe, we wouldn't be living at all. In most, if not all, other universes, the hydrogen that eventually went through a series of nuclear processes to become carbon and other elements would have either been smashed into nonexistence in a cosmic Big Crunch—resembling a black hole in some ways—or dissipated so quickly it never coalesced. Only in a flat, isotropic universe would we live to tell the tale. As Collins and Hawking remarked so eloquently:

The most attractive answer [to the isotropy question] would seem to come from the Dicke-Carter idea that there is a very large number of universes, with all possible combinations of initial data and values of the fundamental constants. In those universes with less than the escape velocity, small density perturbations will not have time to develop into galaxies and stars before the universe re-collapses. In those universes with more than the escape velocity, small density perturbations would still have more than the escape velocity, and so would not form bound systems. It is only in those universes which have very nearly the escape velocity that one could expect galaxies to develop, and we have found that such universes will in general approach isotropy. Since it would seem that the existence of galaxies is a necessary condition for the development of intelligent life, the answer to the question "why is the universe isotropic?" is "because we are here."

By alluding to the notion of a multiverse so soon after DeWitt popularized the Many Worlds Interpretation of quantum mechanics, Collins and Hawking's article brought the concept of parallel universes into even greater focus. They also cast Carter, just starting his career, into the spotlight. Soon he would have much more to say about the idea of selecting our universe from an assortment of options based on its favorability to intelligent life.

THE BEST OF ALL POSSIBLE WORLDS

The notion that the presence of life, especially conscious life, stems from auspicious conditions in the world is an ancient one. According to many religions, the cosmos is designed in such a way as to be favorable to living beings. In the Abrahamic faiths, the Garden of Eden was our first address, and only by defying the landlord were we booted out and exiled to less than ideal quarters. Hence, many believers are hardly surprised that the universe is "fine-tuned" for humankind.

Leibniz's reasoning, that God has a kind of "mental multiverse" in which He weighs various options and chooses from among them "the best of all possible worlds," represents a more sophisticated form of what is called the "argument from design." If one believes in an Intelligent Designer, one might imagine that being poring through blueprints to find just the right one, or simply knowing what would work and what wouldn't. How might a mere mortal fathom the thought processes of an omnipotent immortal?

The scientific process, however, attempts to draw conclusions from testable assumptions and reproducible methods. That's where things get tricky, because what is acceptable science for some represents a matter of faith for others. Consequently, many theorists who dabble in bold speculation treat their research like a bonbon with multiple layers: more conservative, eminently testable conjectures, delectable to the practical-minded and funding agencies, as well as far-flung hypotheses, tasty only to certain specialized palates. For example, Wheeler published verifiable predictions in nuclear and particle physics, as well as lofty speculations about the role of consciousness in the cosmos—each tailored to different audiences.

Carter has similarly contributed numerous important calculations in gravitational physics and other fields, including pivotal results about black holes. Generally, he is rigorous and cautious in his predictions. Nevertheless, an offhand comment he made at the 1970 Clifford Memorial meeting in Princeton about the limitations of the Copernican Principle snowballed into highly philosophical speculations about a multiverse of competing realities. At that conference, he pointed out that, despite Copernicus's notion of the non-centrality of Earth, and the consequent discoveries that the solar system is peripheral in the Milky Way, which is just an average galaxy among many, there is still a role for humankind in cosmology. Our mere existence implies that the cosmos needed to have developed in a manner suitable for the formation of living planets with intelligent beings. Referring to the LNH and other theories about the fine-tuning of cosmological parameters, he expressed agreement with Dicke's idea that they were somehow connected to the development of humans as observers of the universe.

Wheeler, who was in the audience, was listening attentively. Carter's statement was music to his ears, as he had increasingly become fascinated by questions about the role of sentience in physics. He implored Carter to write up his ideas about connections between cosmic conditions and the existence of humanity. Meanwhile, the Collins-Hawking paper on isotropy appeared, giving a shout-out to Carter's (along with Dicke's) speculations.

Given such interest in his musings, Carter decided to write them up in a paper, "Large Number Coincidences and the Anthropic Principle in Cosmology," published in 1974. Taking the position that the standard FLRW cosmology is supported by empirical data, he methodically examined which of its parameters' and other fundamental constants' values might pertain to our own existence.

Carter grouped numerical coincidences, such as large numbers and fine-tuned constants, into three distinct categories. The first are those wholly explained by physical principles and that would be true even in a universe without living beings. For example, because of a delicate balance between inward gravitation and outward pressure, star formation happens only within certain mass ranges. A lifeless universe—which housed stars nonetheless—would still find them developing in similar ways to ours.

The second, which he dubbed the "Weak Anthropic Principle (WAP)," shows how our very existence necessitates the type of region in space and particular era of cosmic history we live in, as well as the range of values of certain physical parameters. For instance, in the history of the Big Bang, it takes billions of years for main sequence (stable, healthy) stars similar to the sun to form. They generally develop in the outer reaches of galaxies, rather than in their compact centers. Furthermore, billions of years in the future, many such stars will have left the main sequence and no longer be able to support life as we know it. Thus, some of the relationships between the current age of the universe, our distance from the heart of the Milky Way, Earth's location in the habitable zone of a solar system with a stable sun, and certain other scientific facts would not require comparisons to other universes to fulfill, but might rather simply have to do with our place in the familiar cosmos as conscious observers.

Finally, most relevant to the concept of a multiverse, is the "Strong Anthropic Principle (SAP)," which requires an ensemble of universes, of which ours is just right for sentient life. Recognizing the speculative nature of a conjecture that requires multiple universes, Carter thought it would be a useful tool, nonetheless, to justify certain large number coincidences—such as the immense weakness of gravitation compared to other forces—that otherwise defied explanation. For example, in universes with a larger gravitational constant, perhaps only giant stars would form that did not support inhabitable planetary systems. If the strong nuclear force were slightly less potent, hydrogen would not be able to fuse into higher elements. In either case, we wouldn't be here. That's why we're in a universe with puny gravitation and a muscular strong nuclear force, rather than the other way around. Our presence explains those relative strengths.

Carter suggested that Everett's Many Worlds Interpretation of quantum mechanics could well provide the ensemble of universes needed to employ the Strong Anthropic Principle. As he wrote:

> According to the Everett doctrine the Universe, or more precisely the state vector of the Universe, has many branches of which only one can be known to any well-defined observer (although all are equally "real"). This doctrine would fit very naturally with the world ensemble philosophy that I have tried to describe.

Dicke concurred with Carter's analysis, writing:

> That the universe is organized as much as it is, and is as uniform and isotropic as it is, might also be an effect of the anthropic principle. If you get the thing [the universe] too badly organized, why I doubt the thing would be very hospitable—[for example], some parts collapsing before other parts get started expanding. . . .
>
> With an ensemble of universes, this is the only kind we could live in.[10]

A popular book by Barrow and American physicist Frank Tipler, *The Anthropic Cosmological Principle*, published in 1986, amplified interest in the idea. Fittingly, Wheeler wrote the forward, as it was his encouragement of imaginative solutions to scientific enigmas that helped inspire thinkers such as Everett, Misner, and Carter to advance far-reaching theories. Applying anthropic reasoning to multiverse ideas soon became a familiar, albeit controversial, component of theoretical physics.

PRINCIPLED MISANTHROPY

By dividing the Anthropic Principle into Weak and Strong versions, Carter significantly broadened its appeal. While the SAP resonates among multiverse mavens and, to some extent, Everett enthusiasts, comfortable with the notion of neighboring realities, those who prefer focusing on the observable universe alone have eschewed it as an untestable gimmick. The WAP, on the other hand, requires no extraordinary assumptions. As a general statement, the notion that the current age of the universe and Earth's location in space link in certain ways to other physical factors such as the efficiency of stellar nucleosynthesis, and therefore key parameters in the world of particles, seems rather uncontroversial. After all, if the sun developed ten billion years ago and was extinct today, we obviously wouldn't be here.

When most physicists address the Anthropic Principle, especially in a critical fashion, they are usually referring to the more far-flung SAP. Pretty much all of the SAP's predictions, such as the relative weakness of gravitation, have been addressed in some manner by other models. That said, sometimes the alternatives involve highly speculative notions—such as large extra dimensions—that similarly have generated controversy.

One leading skeptic of the Anthropic Principle is Roger Penrose, who greatly prefers constructing hypotheses based on the mathematical properties of general relativity. For example, he applied a new constraint to Einstein's hallowed gravitational theory to explain why entropy (a measure of disorder, based in the commonness versus rarity of certain thermodynamic states) began so low in the early universe, allowing it opportunity to rise over time—a necessity for the development of life. Living things operate by turning ordered energy—such as sunlight shining on green-leaved plants and activating their chlorophyll—into waste products. Entropy naturally tends to remain the same or grow. Therefore, it must have been small at the dawn of time. Though the Anthropic Principle might be charged with explaining its initial smallness (otherwise we wouldn't be here), Penrose thought a dynamic reason would be far more effective.

The question of why the cosmos began in a low-entropy state is hardly a new one. In the late nineteenth century, Austrian physicist Ludwig Boltzmann postulated that the entire universe as we know it popped into being—or, alternatively, emerged from a previous stage—through sheer chance in an extraordinarily unlikely state of absolute order and minimal entropy. His reasoning was that, given unlimited time, even the most improbable occurrences are bound to happen. As discussed, in the age of Blanqui and Nietzsche, when the universe was thought to be endless, but with a finite number of components, the notion of unlikely events being eventually bound to occur was compelling. However, knowing that the universe began in a hot

Big Bang and is continuing to expand makes such reasoning far less palatable. Low entropy in the nascent cosmos clearly requires a more cogent explanation.

That's where Penrose saw a golden opportunity, with a proposal called the "Weyl Curvature Hypothesis." It constrains the Weyl tensor—an entity describing certain kinds of geometric distortions, indirectly connected with general relativity—to be identically zero at the time of the Big Bang. Then, it equates entropy with the magnitude of that tensor. (Recall that tensors are mathematical entities that transform in particular ways under rotation.) Subsequently, because entropy starts at zero, it can grow only bigger, but still remain small enough to permit the expenditure of waste energy in orderly processes such as life. With that mathematical trick, presuming one accepts its validity, no anthropic reasoning is needed.

As Penrose remarked in an interview:

> The anthropic principle doesn't do as much for you as you'd like. It doesn't explain the second law of thermodynamics—why the universe was created in a state of such low entropy. On the whole, one finds that the anthropic principle is something you bring in when you haven't got a good theory. People say, "We've got to fix these constants, and the anthropic principle does it for us." It's a way of stopping and not worrying any further.[11]

A specter of Boltzmann's notion of spontaneous emergence of order has returned to haunt the Anthropic Principle. Around 2007, in the context of discussing multiverse ideas, a number of articles appeared debating the notion of "Boltzmann brains": conscious entities randomly emerging from sheer emptiness (or as close as quantum physics allows) via quantum fluctuations. Via pure chance, they might arrive intact with senses, functionality, and perhaps even false memories of a fabricated lifetime. Hypothetically, random disturbances of the void could even bring Blanqui and Nietzsche the comebacks they envisioned in their writings.

Naturally, the chances of single, sentient beings popping out of the vacuum, though immensely improbable at any point in time, would be far greater than the slim likelihood of the emergence of entire, full-fledged universes—given their relative sizes. But the former scenario would create a huge problem for anthropic logic. If even in the meanest, nastiest versions of reality—such as those with high entropy and ample waste, appallingly poor conditions for star formation, and so forth—beings in the form of Boltzmann brains could emerge randomly to observe them (with disgust and despair, no doubt, if they had any semblance of taste), the existence of intelligence wouldn't mandate positive circumstances in the past. If consciousness lacked any connection with cosmic conditions, the Anthropic Principle could not make its cleave between the viable and unviable. That would especially be true if Boltzmann brains were more common than evolved life, as some have speculated.

Note, however, that the existence of Boltzmann brains remains purely hypothetical. Consciousness remains a deep mystery. We don't even know if it could be replicated at all in an artificial way, let alone brought into being through chance quantum events. If Boltzmann brains themselves weigh in on the debate (as talking heads on interview shows, for instance), that's when it would really be time to start worrying. Otherwise, we might reasonably assume, based on human experience, that advanced life requires a semblance of cosmic evolution, and that we wouldn't be here without certain favorable conditions. That is the crux of anthropic reasoning.

Many physicists, while accepting certain aspects of the Anthropic Principle, feel that the term is a misnomer. For example, Bernard Carr, of Queen Mary University of London, has pointed out that the notion pertains to fine-tunings that allow for stars, planets, galaxies, and the chemical elements, rather than humankind specifically.

He noted:

> If you look at what those coincidences are, those fine-tunings are, they've nothing to do with humans in particular. They're general tunings required for things like stars and galaxies . . . and chemistry. I call it a complexity principle, because you could just as well say these tunings are required to have television sets. So, there's really nothing specifically about humans in those fine-tunings. And not all life-forms might be like humans. But whatever your particular life-form, it's still going to require fine-tuning.[12]

In a recent article, accomplished Caltech physicist Stanley Deser—who sadly died in 2023 at the age of 92—weighed the strengths and limitations of the Anthropic Principle:

> One might flippantly dismiss it on the grounds that, like the Holy Roman Empire's name, it is neither Anthropic nor a Principle, but rather a tautology. But despite this perhaps misleading fact, it is shorthand for a particular web of observational data, neither a panacea nor an act of desperation. To summarize, the Principle states the seemingly obvious requirement that the laws of Nature must permit the existence of some sort of sentient life, therefore our chemistry and biology. Like any experimental datum, this one puts stringent limits on our choice of (effective low energy, if not final) laws and their various constants such as masses, charges, Planck's and Newton's, as necessary conditions for organic chemistry, say, to be possible. Thus the word Anthropic is too narrow—natural selection might well have evolved other forms of intelligence, as indeed it did until stumbling on us, and we might well not last long either, but here we are and so must be accounted for!

Given that other intelligent life-forms have yet to be discovered in space, Deser quipped: "Anthropos are so rare, if not unique, in our universe. I call it the Mis-Anthropic Principle: optimal planets and their orbits are forbiddingly unlikely to occur! It in no way invalidates the Principle's usefulness in bounding acceptable physical models; both are necessary."[13]

Returning to Carter's 1974 paper, it is telling that he listed "predictions of the traditional kind," meaning those based purely on physical laws with no reference to the existence of conscious observers, ahead of the WAP and SAP. Indeed, no matter how effective anthropic reasoning might be used to justify certain features of the universe, there will always be researchers who seek physical mechanisms if at all possible.

The Mixmaster universe, though unsuccessful, was an intrepid attempt to solve the horizon problem using the mathematical machinery of general relativity within a single universe, rather than winnowing multiple universes down to one via reasoning. Misner's "traditional prediction" was eminently testable—and ended up falling short in matching the temperature profile of the CMB. There is much to be said for the tried-and-true scientific method, but naturally it doesn't always guarantee progress.

About a decade after Misner's and Carter's respective proposals for resolving various cosmological conundrums, the field was abuzz with an exciting new strategy. In the United States and the Soviet Union, a visionary solution to the horizon and flatness problems emerged involving a brief primordial era of ultra-rapid expansion, called "inflation." Those who wished to avoid thoughts of anthropic reasoning and parallel universes could focus, for a time, on a traditional method involving field theory and solutions to general relativity.

During the 1970s and 1980s, the theoretical physics community was far more optimistic, in general, about standard modern physics methods restricted to a single universe. Successful ventures such as unifying the electromagnetic and weak forces into the electroweak interaction, identifying in collider experiments the exchange particles predicted by that unification, discovering other massive fundamental particles such as the tau lepton and bottom quark, mapping out aspects of the CMBR, and so forth, inspired many theoreticians to believe that representing all the natural interactions via a single theory that makes firm, testable cosmological predictions could well be imminent. Therefore, even though multiverse ideas such as the MWI and SAP—as well as higher-dimensional models such as supergravity and superstring theory—appeared in theoretical journals, that's not where the money was. Funding flowed into projects involving testing ordinary, four-dimensional, single-universe extensions of the Standard Model in particle physics and variations of the Big Bang theory in cosmology.

Modifying the Big Bang to accommodate an extremely brief primordial era of inflation, based on quantum field theory notions connected with the Standard Model, didn't seem, at face value, to carry any philosophical baggage. It didn't smack of science fiction or speculation over pints in a pub. Those who shunned any talk about parallel realities could rest easy.

Ironically, though, that respite would not last long. In short order the concept of "eternal inflation," a natural consequence of a readily triggered inflationary era, implied that our universe is indeed part of a vast multiverse involving the endless formation of bubble universes. Abstract alternative solutions to general relativity such as the Bianchi types were one thing; they either served as part of our own universe (such as in Misner's hopes for a Mixmaster era to smooth out the universe) or they were unphysical and purely mathematical. Pondering other inflating bubbles generating real universes elsewhere in the cosmos was another story.

Soon thereafter, the need to resolve certain seemingly intractable issues in string theory would add to the chorus of those calling for multiverse ideas to be mainstreamed. Those developments would add teeth to the Anthropic Principle, offering it the prospect of a landscape of actual, albeit not directly detectable, competing universes to judge—thereby validating the possibility of selection based on fitness criteria. For many theorists, the SAP applied to a multiverse would become an essential part of the scientific toolkit for characterizing nature. Others, though, would balk at its acceptance, calling for a renewed search for multiverse-free explanations. Fundamental theoretical physics hasn't been the same ever since.

Essentially all inflationary models are eternal. In my opinion this makes inflation very robust: if it starts anywhere, at any time in all of eternity, it produces an infinite number of pocket universes.

—Alan Guth, "Inflation," in
Measuring and Modeling the Universe

Any combination of properties that's physically allowed by the fundamental laws will occur ... an infinite number of times. ... Everything is possible, nothing is preferred. Such is the nature of the multiverse, which is a very nice name for saying what I would call theoretical disaster.

—Paul Steinhardt, AIP oral history interview
by David Zierler, June 2020

CHAPTER FIVE

BURGEONING TRUTHS

Alan Guth, Andrei Linde, and the Inflationary Universe

In nature, sometimes the same ingredients can take on many forms, depending on the environment in which they developed. Carbon gleams as a precious, multifaceted diamond, or blackens as a graphite lubricant. Water presents itself as delicate, gentle snowflakes or harsh, pounding rain. By cranking up and down the dials of temperature and pressure in different manners, adding or omitting agents that seed certain types of growth, exposing to radiation, and altering various other external factors, substances might be made to contract or expand, soften or harden, and go from being brilliantly transparent to mystifyingly opaque. To witness the last transformation, wear a pair of eyeglasses with transitional lenses and take a walk on a sunny day. Phase transitions, as such environmentally sparked structural alterations are called, constitute a ubiquitous part of nature.

It is not far-fetched, therefore, to imagine the universe—or at least parts of it—undergoing a phase transition that alters its composition and hence its rate of growth. Like water and other materials, the substances that fill the cosmos—be they matter or radiation—have their own "equations of state" describing how they respond to pressure differences and other changes. An important difference in cosmology, though, compared to say, ice melting on pavement, is that space itself responds to such transitions by changing its geometry in some fashion.

Thus, in some tens of thousands of years after the Big Bang, the switchover from a radiation-dominated era to a matter-dominated era gradually affected the rate of spatial growth.

As Willem de Sitter demonstrated in his 1917 paper, an empty universe would still grow, provided the version of general relativity under consideration includes a cosmological constant term. In fact, a de Sitter cosmology, consisting of vacuum space fueled by a cosmological constant, would grow exponentially—ballooning much faster than Hubble expansion, the universe's ordinary, gradual growth. An equivalent way of achieving the same aim is positing that space is filled with a substance that has negative pressure. Pack it with a negative pressure ingredient, and it would take off like a firecracker.

In 1981, MIT physicist Alan Guth noted that a primordial phase of exponential expansion might be just the trick for solving the horizon and flatness problems, as well as another dilemma involving why there are no magnetic monopoles (magnets with only north or south poles, but not both) in the universe. His paper, "Inflationary Universe: A Possible Solution to the Horizon and Flatness Problems," galvanized the world of cosmology. It showed how an ultra-brief era of super-rapid stretching would allow for a temperature-evening and geometry-flattening of the observable universe. Anthropic reasoning, many thought at the time, would no longer be needed to resolve the horizon and flatness dilemmas. Through exponential growth during a short sojourn in a super-cooled vacuum state, the cosmos would solve its problems on its own.

AN EXPLOSIVE START

Although Guth wrote about vacuum states, his research did not take place in a vacuum. In the decades leading up to his inflationary universe proposal, great strides had been made in bringing unity to three of the four natural forces: electromagnetism and the weak and strong nuclear interactions. Moreover, various researchers—including, most famously, British physicist Peter Higgs, for whom an important particle is named—demonstrated how scalar fields are essential to completing unification. Through an extraordinary mechanism involving a universal phase transition, the Higgs scalar field lends rest mass to otherwise massless particles. The phenomenon is something like rainwater freezing into thick slush, gripping tires, and causing cars to slow down as if they were burdened with added weight. Without the Higgs mechanism, the particles that make up atoms—such as electrons and the quark constituents of protons and neutrons—would behave like massless photons and thereby soar at the speed of light. In other words, we owe the stability of materials in part to the Higgs mechanism.

What researchers call the "Higgs boson" is the massive remnant of the original scalar field, left over from all of its mass-lending interactions. In our analogy, we might think of it as the innocent piles of slush by the side of the road that did not engage in hindering traffic. Of course, hindering is a good thing for particles that need to be slow to form atoms. Researchers at the Large Hadron Collider first identified it experimentally in 2012 by analyzing proton collision results.

The theoretical discovery of the Higgs mechanism help set physicists thinking about the role of scalar fields in cosmology. Theorists had already determined that adding a scalar field to Einstein's equation of general relativity would have an impact similar to including a cosmological constant: that is, boosting the expansion rate. If no other matter or energy fields were significant, the result of a powerful enough scalar field in the universe would resemble de Sitter's vacuum model displaying exponential expansion over time. De Sitter's odd result, which Einstein had dismissed back in the early twentieth century (1917), turned out to be highly significant in the late twentieth century and beyond. In particular, theorists realized that the Higgs scalar field—or hypothetical scalar fields possibly involved in other particle physics unification mechanisms—could potentially serve as the catalyst for an early, brief interval of exponential growth of the universe—that is, inflation.

The first physicist to posit a primordial era of exponential expansion preceding a far-more-gradual Hubble expansion was Russian theorist Alexei Starobinsky of the Landau Institute for Theoretical Physics. In a way, cosmology was Starobinsky's birthright. He was born in April 1948, the same month that Gamow and Alpher published their seminal paper on how the Big Bang produced the light chemical elements. After becoming interested in how general relativity might be modified at high energies with quantum corrections—pointing the way to a full quantum theory of gravitation—he realized that meant that the universe would behave much differently in its nascent moments. Following the trajectory of a de Sitter model due to the quantum terms acting like a cosmological constant, the very early universe would blow up considerably before settling into familiar, gradual Hubble expansion. Remarkably, he speculated, there might be a way of detecting gravitational waves from the early hyperactive period. Starobinsky published his results, "Spectrum of Relict Gravitational Radiation and the Early State of the Universe," in a Russian-language journal, *Zhurnal Éksperimental'noĭ i Teoreticheskoĭ Fiziki* (*Journal of Experimental and Theoretical Physics*), which was little read in the West at the time, even in English translation. Therefore "Starobinsky inflation," as it later became known, was brought to international attention only after Guth's independent research little more than a year later.

Though they arrived independently at similar conclusions, Guth's background was a world away from Starobinsky's. Born in New Brunswick, New Jersey, in 1947, Guth grew up interested in science, but focused on the very small, rather than the astronomically large. Unusually, he skipped his final year of high school, began university studies at MIT in 1964 at the age of seventeen, and remained there until he completed his PhD in Physics in 1971. His dissertation, under the supervision of Francis Low, concerned how quarks are confined to become elementary particles. After nine years of postdoctoral and other academic positions at various institutions including Princeton, Columbia, Cornell, and the Stanford Linear Accelerator Center, he returned to MIT as an associate professor of physics in September 1980. By then, he had completed and submitted his paper on inflation, which was published the following January.

Before tackling cosmological questions, Guth's focus was on GUTs (grand unification theories) intended to meld electromagnetism, the weak force, and the strong force into a unified interaction at ultra-high energies. In the 1960s, American physicist Steven Weinberg, Pakistani physicist Abdus Salam, and American physicist Sheldon Glashow had already contributed the elements of a completely unified model of electromagnetism and the weak force, called the electroweak interaction. Electroweak unification served as the blueprint for the Standard Model of particle physics, which also includes the strong force. The latter interaction is modeled with exchange particles, also known as force carriers, called gluons volleyed between quark components. Additionally, in the Standard Model, the Higgs mechanism supplies rest mass for many of the particles. GUTs took this idea one step further by attempting to show how a single set of exchange particles could have divided into photons (the carriers of electromagnetism), W and Z bosons (the carriers of the weak force), and gluons. Researchers tried to find a single mathematical group that included all three. Implications would include proton decay and a set of new particles. However, efforts so far to find a unified group theory have been to little avail.

In 1978, Guth attended a talk by Dicke about many of the outstanding conundrums in cosmology, including the flatness problem. He left Dicke's talk with a newfound interest in applying the apparatus of particle physics, with which he was familiar, to resolving cosmological questions. By late 1979 he realized that the phase transition common to GUT models as temperatures dropped in the early universe could lead to an epoch of exponential expansion. "Spectacular Realization," he wrote in his research notebook, boxing in those words twice for emphasis.[1] Undoubtedly because price rises were on everyone's mind in those times, he dubbed that cosmological period "inflation."

SUPER COOL MOMENTS

Under normal circumstances, liquid water, if cooled down below its ordinary freezing point of 0 degrees C (32 degrees F), will harden into solid ice. However, nature's rules sometimes have noteworthy exceptions. In a process called "super-cooling," if water lacks impurities and is cooled down very slowly and carefully, it can maintain its liquid state at sub-freezing temperatures.

In melting a block of ice, a certain measure of energy must be added, called the "latent heat of fusion." Conversely, in freezing liquid water into ice, that latent heat is released. In the case of super-cooling, however, yet another possibility transpires. The latent heat is stored for the time being, as long as the water remains isolated and in the super-cooled state. Eventually, though, the latent heat is released into the environment, and the water freezes to some extent, typically becoming a slush.

Guth knew about an analogous process in the spontaneous symmetry breaking of GUT models. Spontaneous symmetry breaking refers to how a system with perfectly symmetric rules might randomly drop into a state that breaks one or more of its symmetries. For example, imagine a spinning top with four faces resting precariously on a pivot. As it is spinning, its faces whirl by at an equal height, presenting a perfect example of rotational symmetry. Because, while the top is spinning, a point on its rim is equally likely to be in any angular direction, we also say that it has an angular degree of freedom. Suddenly, due to a random push—or in quantum mechanics, a random transition that might take place as environmental temperature is lowered—it tips over, and rests on one of its faces. The top, although it remains the same symmetric shape, no longer exhibits the same rotational symmetry. By not spinning and simply resting, it loses its angular degree of freedom. In other words, by falling on one of its sides, a particular direction is favored, breaking the symmetry.

Losing a degree of freedom is one way of characterizing a phase transition. Liquid water forgoes its freedom to glide, for instance, when it turns into ice. Just as in freezing, the spontaneous symmetry breaking of a GUT releases a certain amount of latent heat. Moreover, in a manner similar to slowly lowering the temperature of pure water, the GUT might undergo super-cooling to a false vacuum state, hoarding its latent heat, before transitioning to the true vacuum, releasing that energy in the process. A false vacuum is a metastable (temporarily stable but eventually decaying) state that has a higher energy level than the true vacuum, in which a particle physics entity, such as a scalar field, might lodge for some time.

In exposing super-cooled water to a cold environment, pockets of liquid water might remain as other segments lose their latent heat and become ice, leading to a kind of slush. Similarly, in the GUT analogy, sectors of false vacuum might persist among the true vacuum. Because the false vacuum would be at a higher energy than the true vacuum, it would have a scalar field that behaved something like a cosmological constant, causing exponential expansion in line with a de Sitter model. That expansion would cease once the false vacuum decayed into true vacuum, via "quantum tunneling": random quantum leaps allowing passage through otherwise impenetrable barriers. That tunneling would act as a release value, similar to how party balloons eventually deflate, even if they aren't popped, by means of inevitable minor leaks. Without its driving energy, inflation would end. Subsequently, in a process called "reheating," the stored latent heat would release and transform into a flood of particles.

Now imagine that the universe started off in a chaotic, gradually expanding state, full of energy fields with the symmetries of a GUT. Some 10^{-35} seconds after it originated, its temperature drops, and sectors super-cool into false vacuum bubbles. Those bubbles have a suitable scalar field and thereby burst into exponential mode, blowing up by a factor of 10^{25}, before they decay into true vacuum. A segment tinier than a proton would suddenly expand to the size of a baseball. (Following inflation, that baseball-sized region would, over billions of years of much slower growth, expand to the size of the observable universe.)

A brief inflationary era, fractions of a second after the birth of the universe, would propel close regions that have equalized their temperatures extremely far away from each other. That is, Cosmic Microwave Background Radiation signals from the opposite sides of the sky today would have been so close together before the brief inflationary era that they would surely have been in thermal contact. Imagine if an evenly heated kettle of broth were suddenly stretched in a fraction of a second until it was the size of the Milky Way. It would still have an even temperature. Similarly, the snap expansion would maintain near-uniform temperatures of space, even as points are flung far apart. Hence, an inflationary era would solve the horizon problem.

A brief exponential burst would also resolve the flatness problem. The extraordinary expansion would flatten that part of space, similar to pulling on a bedsheet from all sides with a quick tug. All wrinkles in what ended up being the observable universe would be gone.

Finally, a problem with excess monopole production in GUT models would be resolved. If inflation took place during or after that creation of single-poled magnets their density would drop significantly, rendering them rarer than an Australian kangaroo one-ton gold coin. The current lack of monopoles would be well explained.

Guth hoped, in the end stage of his model, that all inhomogeneous, irregular regions in space would have been pushed far away by the inflating, false vacuum bubbles. Theoretically, those false bubbles would exude into the true vacuum through quantum tunneling and vanish. From that point forward, the universe would gradually grow as a FLRW model, following the observed Hubble expansion.

However, that's not how his model worked. In what was called the "graceful exit problem," quantum tunneling was not efficient and reliable enough to carry out the process of ending inflation. As a result, the demise of the false vacuum bubbles would be inconsistent, leading to a frothy inhomogeneous state, much like the slush sometimes produced after super-cooling water, rather than the near-isotropy we see in the CMBR sky maps today. Guth conceded that his model was incomplete, and required more work in figuring out how to turn off inflation.

Fortunately, researchers in the Soviet Union and the United States were up to the task. In Moscow, Andrei Linde developed a different method for triggering and ending inflation involving a scalar field with a flat potential energy curve. Potential energy curves are something like roller-coaster tracks. Just as a ride car could coast for a long time on a level track, in comparison to a steep drop, similarly a scalar field could persist a while on a flat potential energy curve. The field would remain in the energetic state for a time, triggering exponential expansion, until it dropped into a true vacuum state of (relatively) zero energy. Reheating would occur, with a flood of particles released into the vacuum. From that point forward the universe would follow a much-more-subdued Hubble expansion. That notion, called "new inflation," avoided the graceful exit problem because no quantum tunneling was involved. Independently, American physicist Paul Steinhardt and his student Andreas Albrecht, working at the University of Pennsylvania, arrived at a similar concept around the same time. Therefore, all three researchers are recognized for the discovery of new inflation.

The biggest issue with new inflation was "fine-tuning." It was hard to justify why the potential energy curve would have exactly the right shape for the scalar field to persist at a certain energy level, do its job in triggering just the right amount of inflation, and then transition to the vacuum. Researchers, such as Linde and Steinhardt, continued to look for alternatives that would be more natural.

COSMIC KALEIDOSCOPE

In the summer of 1982, the Nuffield Workshop on the Very Early Universe, held in Cambridge, England, resounded with a chorus of excitement and activity concerning the possibilities and opportunities of inflation. With the graceful exit issue resolved, "new inflation" seemed a very promising way forward in cosmology—along with any similar models that avoided the need for tunneling and thereby weren't stymied by slow, uneven transitions to FLRW behavior. The universe could start out absolutely chaotic, switch into exponential expansion, and emerge as smooth as silk.

Hawking, who had already become an internationally respected figure in physics due to his pivotal contributions to astrophysics and general relativity, and was soon to enjoy global fame as a popularizer as well, was, if anyone, the star of the show—among many extremely talented individuals. His imprimatur meant much to the attendees. Regarding inflation, arguably his support was important to establishing it quickly as the mainstream modification to the original Big Bang theory.

Aside from resolving the horizon and flatness problems, one of the extraordinary outcomes of having an inflationary era in the very early universe was the massive production of elementary particles upon reheating at the end. No longer could the Big Bang be criticized (as it was by British astrophysicist Fred Hoyle and others) for purporting that all the matter and energy in space burst out of sheer nothingness in the opening gavel of time—thus violating the law of conservation of mass and energy in a most egregious manner. Rather, the energy stored by the inflating bubble in its sudden outburst would be put to good use upon the close of that episode in bringing the universe all the particles that we know and need.

Moreover, enough particles would be created to resolve one of the quandaries raised by Dirac in his Large Numbers Hypothesis: Why are there some 10^{80} (later revised to 10^{90}) particles in the observable universe? Rather than connecting principally with the strength of gravitation and other parameters, as Dirac surmised, the number stems directly from the length and scope of the inflationary epoch. It would store enough energy during that primordial phase of exponential expansion to release that enormous torrent of particles once it was over. Because that aspect of the LNH would be explained, the Anthropic Principle and multiverse concept would not be needed for justification. There would be one universe, many researchers hoped, with one primeval burst of inflation, lasting for just the right amount of time to justify the nearly consistent temperatures of the CMBR, flatten the observable universe, dilute its monopoles beyond ready detection, and explain the huge bounty of elementary particles and their incredible amounts of mass and energy. It was a tall task for sure, but it seemed that inflation, remarkably enough, was up for the job.

There was yet another trick up inflation's sleeves. If matter and energy were distributed absolutely evenly in the very early universe, it wouldn't harbor the seeds of higher density needed to trigger eventual gravitational collapse of gas clouds into stars and planetary systems. That is, pure sameness in the very beginning wouldn't lead to the diverse cosmos seen today—with its stunning formations, as revealed by the James Webb and Hubble space telescopes. Happily, as shown by researchers at the Nuffield Workshop, including Guth, exponential expansion magnifies any quantum fluctuations that develop randomly during the inflationary epoch. The uncertainty associated with quantum mechanics of certain physical parameters, such as energy values, at tiny scales would produce chance microscopic variations in the energy density of the primordial universe. Inflation would then blow up these minute, random fluctuations into sizable variations in the density of energy and—as new particles emerged—matter. The higher-density clumps of material would gradually form the seeds of even larger chunks—coalescing, over the eons, into massive enough clusters to coagulate through gravitation into the first stars and, eventually, galaxies.

Once inflation ends, the emergence of the observational universe into much-slower growth stamps the matter and energy released with the imprint of those fluctuations. As Guth demonstrated in a subsequent paper with South Korean physicist So-Young Pi, "Fluctuations in the New Inflationary Universe," those variations would display a distinct scale-invariant imprint. In other words, correlations between bumps would occur with roughly equal magnitude on all scales. Because such density fluctuations would lead to temperature variations in the CMBR, inflation thereby predicts a telltale signature in the radio sky. Such minute scale-invariant fluctuations were picked up in subsequent decades by the COBE, WMAP, and Planck Satellites. Many physicists viewed such detection as the "smoking gun" for inflation, justifying its widespread acceptance.

At Nuffield, high hopes were tempered, though, by the need to fine-tune inflation. New inflation, with its flat potential energy curve, leading to a rollover at the end, seemed artificially created to generate the correct results without a clear physical justification behind it. The scalar field most widely discussed, the Higgs field, had not been shown, through field theory calculations, to have that precise profile. A more generic "inflaton" field could be matched to the right specifications, but that required making very specific assumptions.

One answer would be to invoke the Anthropic Principle in a multiverse of possible scalar field potential energy curves. Only in a region in which a scalar field's dynamics displayed the correctly shaped curve would inflation begin in a way that led to a universe that eventually supported at least one planet with intelligent life and would we be able to say "Hey, that's us!" Hawking, Linde, Barrow (who later co-authored with Frank Tipler a noted book about the Anthropic Principle), and some of the other participants were comfortable with that option. Linde sought to develop that notion further.

As Linde recalled:

During the conference in Cambridge, I had written a paper which described two different ideas—first, that you might solve the singularity problem, and then that you also have this anthropic landscape, so to speak. . . . Then on the flight back, Starobinsky and [Igor] Novikov, who was there also, told me that the idea about absence of singularity was wrong. . . . So, I withdrew the paper from *Physics Letters*. It was a painful decision and probably a stupid one because this was the first paper describing what was later called inflationary multiverse.

However, we were writing proceedings for this conference, and for me, proceedings of the conference in Cambridge means that everybody would know about it, okay? So, who cares about *Physics Letters*? This is the Proceedings of this famous conference, the best in my life. I had written there a large description of inflationary theory and ended up with this discussion saying that it does not solve the singularity problem, but it allows you to make a scientific justification for the anthropic principle because it gives you an option in which universe to live.[2]

In contrast to those fine with anthropic reasoning, other Nuffield attendees, notably Steinhardt, considered inflation to be a way of avoiding talk of a multiverse and leaving humans out of the equation. However, at the close of his own talk, he brought up an issue that he thought was minor but turned out to be a big deal. Quantum fluctuations could reignite inflation, making it keep going in other regions.

As Steinhardt remembered:

That idea, when I wrote about it in the conference proceedings, was sort of the beginning of the idea of an eternal inflation that leads to a multiverse, which is the first big huge crack in the inflationary paradigm. But it was not recognized as such at the time. Not even by me. I thought I was identifying a feature of the theory—not highlighting a fatal flaw that would eventually come to be known by the community and that remains with us today.[3]

Linde was similarly intrigued by that idea, but he didn't consider it such a bad thing. It just meant that inflation was easier to trigger. Consequently, the smoothness and flatness of the observable universe would be straightforward to generate. The fact that such a snap triggering mechanism would generate countless seeds of expansion and likely create a multiverse, he didn't see as a dealbreaker.

ANYTHING IS POSSIBLE

Linde continued after the conference to try to resolve the fine-tuning problem by looking at scalar field potentials. At one point, he engaged in discussions with fellow Soviet physicist Alexander Vilenkin, born in Kharkiv, Ukraine, who had immigrated to the United States and was on the faculty of Tufts University. Vilenkin was independently pursuing a somewhat different avenue—triggering inflation by means of quantum fluctuations of the vacuum.[4] His 1982 paper "Creation of Universes from Nothing" attempted to avoid the need for a Big Bang altogether, replacing it with a bubble bath of quantum-induced centers of growth, leading to different universes. In other words, it was a multiverse from which our enclave originated, but many others were out there as well. Vilenkin showed how Steinhardt's insight was on the mark—random quantum perturbations could easily spark epochs of exponential growth.

Meanwhile, Linde's exploration of scalar field potentials led him to an astonishing discovery. The carefully crafted flat potential of the new inflationary scenario was not needed. Rather, very simple potential energy curves could readily do the job. One such curve, commonly discussed in basic university physics courses, is the rise and fall in potential energy of a spring (or rubber band, and anything elastic) as it is stretched and released. Called a "harmonic oscillator" potential, it takes the form of a parabola. A scalar field rolling over such a potential curve would stimulate a bout of inflation. Many other simple curves could do the task too. Therefore, in a model Linde called "chaotic inflation," the very early universe could exhibit very general kinds of energetic transitions (due to symmetry breaking in its fledgling moments, for example) and still undergo sufficient inflation to smooth itself out.

As Linde pointed out:

> That is why I invented chaotic inflation, which was based on the simplest set of theories with the simplest potentials ever, like a potential of a harmonic oscillator, and just by magic this worked, there was inflation in the early universe. . . . So, there was a huge class of models where one can have inflation. . . . And if the universe consisted of many different parts, then in some of them inflation may not start, and in some other it may begin. But the parts which do not inflate remain small and irrelevant, and the parts that inflate become large and uniform. One can create order out of chaos. That is why I called it chaotic inflation.[5]

By the mid-to-late 1980s, the work of Linde on generic potentials and of Vilenkin on quantum fluctuations converged to reveal a finding that confirmed Steinhardt's worst fears. Rather than there being a single universal history that includes a unique inflationary epoch, such phenomena are ubiquitous and ever present. In what Linde called "eternal inflation," inflationary outbursts are guaranteed to happen again and again, stimulated by quantum events that induce the commonplace energy profile that creates exponential expansion. That meant not just a single universe, but an endless sea of bubble universes, each growing into a different reality. Some of the bubbles would develop into successful, long-lasting universes, such as ours; others would die out very rapidly.

In eternal inflation, the observable universe would have innumerable neighbors. However, like the vastest imaginable sprawling suburb, our acreage would be so immense due to its exponential expansion in the past that we couldn't possibly visit them or observe them at the present moment.

With the realization that inflation is eternal, Steinhardt would eventually abandon the theory he helped create. It did not lead to the unique universal chronicle he was seeking. Rather, it produced a multiplicity of stories with no clear resolution—like a film with numerous vignettes but no overriding plot, or the ultimate cacophonous attempt at a symphony. Harmony seemed to be lost, in his view, in favor of sheer pandemonium.

As Steinhardt explained why he came to lose hope in inflation:

> Just imagine—your theory was supposed to explain why things are this way and not that way.... Instead, what you find is—your theory predicts that all are equally possible.... Fluctuations that you would say would be rare and unlikely, but that might keep inflation going, become huge compared to the regions that you thought were typical. That is how the simple classical universe that we originally thought inflation produces—and that makes definite predictions—turns into the multiverse that was not anticipated and that makes no predictions.[6]

Steinhardt came to believe that the inflationary model was not true science, because, in his view, it was not falsifiable. Once theorists demonstrated that chaotic inflation—representing a broad sweep of possibilities—turned out to be eternal inflation, we became faced with an endless array of bubble universes beyond our reach. Each would have its own conditions today that we couldn't possibly test at present. Numerous other realities could harbor the same physical parameters as ours—or they might not. We wouldn't know. No telescope might reach beyond the horizon of our observable universe to search for the properties of other parallel spaces. Therefore, in Steinhardt's opinion, belief in inflation has become a matter of dogma, faith, and stubborn adherence to an outdated concept, rather than genuine scientific inquiry.

SCARS OF THE EPIC COSMIC BUBBLE BATTLE

Despite the skepticism of eternal-inflation critics, it is conceivable that the notion is a testable proposition—not based on present-day telescope observations, but rather by searching for imprints of bubble collisions from the distant past in the CMBR. If a neighboring bubble impacted our inflating bubble in the fledgling instants after the Big Bang, like smashing volleys fired by conflicting armies, perhaps the bumping would leave a subtle impact in the radio sky. The wound might manifest itself as an inhomogeneity in the energy distribution of the developing observable universe. Perhaps it didn't completely heal and remained as a subtle flaw in the slowly cooling energetic background. Long after wars are over and troops are dispersed, battle scars can serve as reminders of times of combat. Similarly, bubble scars in the CMBR might indicate early cosmic clashes in a multiverse arena.

The notion of searching for signs of colliding bubbles in cosmic microwave background data collected by the WMAP satellite instrument came about when Hiranya Peiris, a member of the research collaboration associated with that probe and a professor at University College London, co-organized a summer program in 2009 at the Aspen Center for Physics, nestled in Colorado's scenic mountains.[7] She struck up a cosmological conversation with Matthew Johnson

Image 16. Sri Lankan physicist Hiranya Peiris, part of the 2017 Breakthrough Prize–winning WMAP science team, who has helped design and conduct tests of eternal inflation and other cosmological hypotheses. Credit: Photographed by Niklas Björling for Stockholm University and included with his permission; courtesy of University College London.

of the Perimeter Institute, based in Canada in Waterloo, Ontario. Johnson suggested the idea of testing eternal inflation by looking for primordial bubble collisions in the wealth of information collected by WMAP over a seven-year span. Agreeing to collaborate, the two developed mathematical models of the shock waves that would result from such hypothetical crashes. Such cataclysmic events would produce telltale spots, like impact craters due to astral collisions. Those spots would have a recognizable profile with a particular kind of symmetry associated with the bumping together of two bubbles and relatively long-range correlations expected from the subsequent stretching due to inflation.

Such an ambitious and meticulously conceived project was, in part, an emblem of Peiris's long-standing fascination with the patterns of the heavens. As she recounted, she became interested in astrophysics and cosmology through "a combination of the dark starry skies of my native country Sri Lanka, and encountering Carl Sagan's *Cosmos* and Stephen Hawking's *A Brief History of Time* at an early age."[8]

Above University College London, the urban sky—over one of the most populated cities in the world—is not so dark. However, thanks to Hawking, Penrose, Martin Rees, and many other cosmologists who've carried on the legacy of their late, brilliant mentor, Dennis Sciama, sweeping visions of the universe have flourished in the UK. In the land where the 1919 solar eclipse expeditions that tested general relativity were planned and their results analyzed and announced, imaginative ways of probing the very limits of reality have become a time-honored scientific tradition. Gazing far beyond the realms explored by her predecessors, Peiris, in tandem with Johnson and others, has extended that pursuit by engaging in a bold quest for evidence of other domains in an unimaginably vast multiverse.

Unfortunately, the WMAP seven-year results proved disappointing for bubble collision hunting. In the entire CMBR sky, there were no statistically significant examples of the kind of symmetric patch the team—including Peiris, Johnson, and others—was seeking. More hopefully, there were four candidate examples that stood out somewhat from the noise and deemed worth further analysis via data from the Planck satellite. Planck was higher resolution than WMAP, lending hope that the spots would be more prominently distinct from the background. Alas, that was not to be. The impact-zone candidates maintained about the same distinctiveness. It would be like an island castaway scanning the horizon for potential rescue ships, and seeing four vague shapes, picking up binoculars, searching again and seeing the same blotches. In other words, the prospects of signs from beyond remained inconclusive.

To advance with the search, the team realized that it needed even more powerful tools and methods. A novel idea emerged to use the polarization of the cosmic background radiation as a means of testing for collisions. Light is based on electric and magnet components, perpendicular to each other as they move through space, that can be analyzed by looking head-on at clockwise and counterclockwise circular motions. If clockwise and counterclockwise are equally mixed, the signals are said to be unpolarized. However, if there is a bias, the light has polarization, which potentially can be measured. If there are bubble impacts, that leads to a predictable profile.

In 2014, the scientific community was abuzz with the announcement of purported evidence for inflation in CMBR polarization results, gathered by researchers working on a different project. Rather than taking advantage of the cold, empty vacuum of space—ideal, but expensive and risky—the BICEP 2 (Background Imaging of Cosmic Extragalactic Polarization 2) team established its detector in the cold, empty continent of Antarctica near the South Pole.

"The South Pole is the closest you can get to space and still be on the ground," said team leader John M. Kovac of the Harvard-Smithsonian Center for Astrophysics.[9]

The researchers were specifically looking for a phenomenon called "B mode polarization": curly patterns in the signals, that theoretically could be produced by the impact of primordial gravitational waves from inflation's burst. They were searching for evidence of inflation in general, rather than signs of more exotic phenomenon such as bubble collisions. After collecting data for some time, and trying to sort the signals from the noise, they felt confident enough in March 2014 to announce success.

Dramatically, team member Chao-Lin Kuo, a colleague of Linde's at Stanford (where Linde had been appointed to a professorship), knocked on his door, bringing a bottle of champagne to celebrate the news. Linde, who was about to go on vacation in the Caribbean with his wife Renata Kallosh (also an eminent physicist), was floored. Guth, informed around that time too, was similarly excited. They had not thought such a direct test of inflation's impact on the early universe was possible. The results seemed absolutely miraculous and life-changing.

"Space Ripples Reveal Big Bang's Smoking Gun," proclaimed a *New York Times* headline.[10]

Sadly, further analysis of the results showed that the team had spoken far too soon. As Planck satellite data indicated, the curly patterns could be fully explained by noise due to galactic dust. After recovering from the shock and disappointment that their ostensibly world-changing findings literally vanished into dust, the team has upgraded their equipment and continued with the project.

Neither evidence of primeval cosmic crashes nor of inflation itself in the CMBR sky through polarization analysis seem soon forthcoming. Peiris recently described the current state of finding bubble collision scars: "I am afraid it is stuck waiting for much high quality data on the polarisation of the CMB over a large portion of the sky. Having such data will allow us to distinguish the signal from sources of confusion. This data may soon arrive however, from the Simons Observatory on the ground and the LiteBIRD satellite in space."[11]

When asked further about the current prospects of testing eternal inflation, Peiris remarked:

> I think the chances are slim because nature has to be kind to us to see a signal. But these tests should be done, because the implications of the theory are so momentous. There may be another avenue to test the physical understanding of the bubble nucleation process itself through a completely different method—in the lab, within an analogue quantum simulator. I am part of a consortium which is taking steps towards realising such an experiment.[12]

The Quantum Simulator for Fundamental Physics (qSimFP) consortium, to which Peiris along with other investigators from seven research institutions in the UK and five international partners belong, aims to use quantum technology to simulate conditions in the very early universe. Perhaps those simulations will demonstrate how quantum fluctuations impact cosmology in a manner that would resolve once and for all whether or not inflation must be eternal and lead to a multiverse.

If those sophisticated models show that inflation absolutely stirs up an endless cascade of bubble universes, as Linde, Vilenkin, Guth, Steinhardt, and others surmised, the debate over whether or not to pursue alternatives would become even more intense. Steinhardt has already jumped ship, thinking that a lack of predictability due to an infinite maze of vying realities should sink the theory. On the other hand, the inflationary model's prediction of scale-invariant temperature fluctuations has been stunningly reproduced in increasingly precise CMBR data, leading many to hold tight in the only solid vessel we have. A vast, largely unobservable multiverse could well become the necessary consequence of an overwhelmingly supported theory that otherwise matches all known data.

As cosmologists try to fathom the unimaginably vast, high-energy theorists and experimentalists attempt to comprehend the extraordinary small: the realm of subatomic constituents, from quarks and leptons to exchange particles and the Higgs boson. Increasingly, physicists realize that there is a profound connection between the two extremes.

In recent decades, the age-old quest for unity in the realm of subatomic entities and natural interactions has led to an astonishing link with the multiverse of bubble universes in eternal inflationary cosmology. The "string landscape" idea purports to show how the Anthropic Principle, applied to a vast menagerie of universes distinguished by their curled-up, higher-dimensional internal spaces, singles out the realm with the most favorable physical parameters, such as a small cosmological constant. Nobel laureate Steven Weinberg's advocacy of the need for a multiverse, filtered via the Anthropic Principle to leave only beneficial outcomes, promoted the idea to the mainstream physics community in unprecedented fashion. Ideas from across physics would be combined: decades of outrageous unification schemes stemming from Kaluza-Klein theory; the labyrinthine structures of Wheeler's quantum foam and Everett's many worlds; the strangest consequence of modern cosmology, an endless sea of inflating bubble universes; one of the most provocative methods for explaining nature's laws, the Anthropic Principle; and finally the much-debated replacement of elementary particles with energetic strings living in high dimensions. They all seemed to culminate in the bewildering notion of an enormously intricate string landscape. The string multiverse, the child of a veritable Addams Family of weird twentieth-century ideas, could possibly lead to profound revelations about the world around us. Alternatively, if it fails to achieve its goals in a way that satisfies the mainstream physics community, and cedes not even an iota of indirect physical proof, let alone direct evidence, it might take us to even stranger theories.

Recall Kipling's *Just So Stories*, such as "How the Leopard Got Its Spots?" The story helps us remember that leopards have spots. The same might be true of "Just So Stories" in physical science. Maybe the absolute value of the cosmological constant is accounted for by the selection of a habitable universe from a multiverse. Or maybe a deeper physical theory will account for it without recourse to a multiverse.

—P. James E. Peebles, personal communication to the author, May 2022

CHAPTER SIX

TANGLED UP IN STRINGS

Edward Witten, Steven Weinberg, and the Higher-Dimensional Landscape

In little more than a century since Einstein, de Sitter, Friedmann, and others introduced the first scientific cosmologies, our vision of reality has profoundly enlarged to encompass many aspects that cannot fully be measured, if at all. A vast, four-dimensional universe housing innumerable galaxies has vaulted from fanciful speculation to widely accepted truth. We recognize the existence of dark matter and dark energy—unknown cosmic components that scientists are vigorously trying to identify. Because the speed of light constrains the range of telescopes, we need to grapple with the likelihood of enormous—and perhaps infinite—reaches far beyond the horizon of observability.

In quantum physics, there is similarly much that is fundamentally unknown about nature at any point in time. But rather than pertaining to deep space, it directly affects us. Much is hidden in the innumerable filing cabinets of Hilbert space (the unlimited mathematical domain in which quantum states evolve), for which only certain drawers containing precise information, such as the exact position of a particle, might open at the same time as others. While Einstein, David Bohm, and others sought ways of restoring local realism (seamlessly connected objective reality) to physics by positing unseen mechanisms working behind the scenes to bridge the gaps, quantum computation and quantum measurement theory show how uncertainty is fundamental.

Nonlocality and blurring of possibilities have become widely accepted. Long-range correlations of particle properties, through what is called "quantum entanglement," have been borne out by numerous, meticulously crafted experiments. What remains subject to debate is largely if and how collapse occurs—with the Everettians stressing the continuity of a Many Worlds multiverse and others seeking physical mechanisms. However, even without the hypothesis of splitting into alternative realities, quantum physics, with its entangled web of correlations and connections, is plenty bizarre and labyrinthine enough.

Top off all that strangeness with the distinct possibility that our universe is one of countless bubble universes—as predicted by eternal inflation. Truly, the going is getting weirder and weirder. Yet, if the mainstream community is willing to stomach four dimensions; vast, unobservable regions of the cosmos; unseen substances that make up the majority of all stuff; and an intricate Hilbert space, one might as well shrug one's shoulders and accept a bubbly multiverse as well—perhaps even toasting it with effervescent champagne. Traditional, mundane objects with their simple Newtonian laws seem so nineteenth century, in comparison, like horse carts and gas lamps. We've moved so far beyond the eminently measurable that it is hard to imagine turning back.

Curiouser and curiouser modern physics has become as it recedes from the tangible and embraces an abstract wonderland. String landscapes, in which universes with differently curled, six-dimensional compactified spaces engage in a survival of the fittest, seems in many ways absolutely baffling. Yet perhaps it is the inevitable offspring of many of the ideas discussed, from a universal wave function and quantum foam to Kaluza-Klein theories, the Anthropic Principle, and eternal inflation. Those radical notions form a hall of precedents leading to a theoretical physics approach that, faced with the elegant simplicity of the Standard Model of the world of particles and the astonishing uniformity and other favorable conditions of the expanding universe, attempts to distill them from the mind-boggling complexity of a multiverse, this time built on the myriad ways of compactifying the extra dimensions of string and M-theories.

When Kasner's nephew suggested the concept of a googol—ten raised to the hundredth power, represented by the numeral 1 followed by 100 zeroes—the idea of a physical application for such an enormous quantity seemed remote. Now, string theorists confidently talk about manifolds of ten or eleven dimensions, with six of them curled up like twine into Planck-scale balls—in some 10^{500} (1 followed by 500 zeroes—or a googol to the fifth power) possible configurations. Of these, theorists believe that only limited possibilities represent the Standard Model—with its electroweak and strong interactions, quarks, leptons, and exchange particles—and the rest are alternative realities with radically different kinds of particles and interactions. Each would also be identified with a distinct value of the cosmological constant. Without selection rules, narrowing those vacua down from 10^{500} to 1 would be a tall order indeed—harder than filling the observable universe with coins and searching through the entirety for an ultra-rare penny. Following a suggestion by the late Nobel laureate Steven Weinberg, some hope that the Strong Anthropic Principle would be able to do the sorting and weeding out—singling out the one possibility that leads to intelligent life. Others are dubious that it would be a robust-enough selection mechanism to handle the job.

RUBBER BAND LAND

String theory started out quite humbly, in the early 1970s, as an attempt to show how the strong nuclear force connects quarks and antiquarks into pairs to form the broad category of elementary particles called "hadrons." Hadrons, any particles that feel the strong force and are thereby made of quarks and antiquarks, include mesons (quark-antiquark pairs) and baryons (three quarks or antiquarks each) as subcategories. Baryons, in turn, consist of protons, neutrons, and more massive subatomic constituents.

It took some time to find the missing theoretical ingredient to keep quarks confined to pairs and triplets, as they are in nature. Eventually the notion of gluons was cemented into hadron theory as the adhesive intermediaries between quarks. Until then, physicists pored through scattering data trying to find a hypothesis reasonable enough to stick. Among those was a formula discovered by CERN physicist Gabriele Veneziano, connected with an approach called "dual resonance theory" that nicely predicted scattering profiles involving hadron interactions. Though a formula is nice, a physical analogy is even better, and that's why string theory was first developed.

Some theories begin with a lab accident, but string theory started with a car failure. In summer 1970, physicist Yoichiro Nambu of the University of Chicago wrote up lecture notes on what he called "hadrodynamics"—including the notion of flexible strings connecting quarks and antiquarks to model the strong interaction—for a planned talk at a symposium on higher physics to be held in Copenhagen that August. He mailed the manuscript to the symposium organizers in advance. All set to fly to Europe, he decided to drive to California and drop the rest of his family off with friends, so they'd have a nice place to stay in his absence. While crossing the Great Salt Lake Desert in Utah, Nambu's car broke down. He and his family were stranded in a remote town for three days while it was repaired. By the time it was drivable, he had missed his flight. Rather than rebooking and rushing off to Denmark, potentially arriving too late to deliver his talk, he decided to skip the symposium and go on vacation in California with his family instead. While his lecture was never delivered, and did not appear in any conference proceedings, luckily his manuscript was preserved and eventually published. In it, he sketched out ideas for connecting quark-antiquark pairs with elastic strings or rubber bands in order to model how they behave. Applying quantum rules to the strings' vibration modes produced patterns matching dual resonance theory predictions. In the meanwhile, other researchers such as Holger Nielsen and Leonard Susskind had independently arrived at a similar notion of conveying the strong force through strings—modeling the surprisingly narrow range of that interaction, on the scale of an atomic nucleus, as well as its high potency in that tiny region.

Soon, physicist Claud Lovelace of Rutgers University made a startling discovery about the mathematical underpinnings of hadronic string theory. In the standard four dimensions of space-time, certain unsavory faster-than-light terms, called "tachyonic cuts," pop up in the equations. As Einstein's special theory forbids such expressions, Lovelace sought a way to rectify the situation. Curiously, he determined that raising the number of dimensions to twenty-six served to eliminate those troublesome terms. Finding that very weird and hard to take seriously, he mentioned his result offhand at a Princeton University seminar he was invited to deliver. Upon hearing about a world of twenty-six dimensions, the audience burst into loud laughter.[1]

While, for the early 1970s physics community, a dimensionality of twenty-six seemed ridiculously large, extending general relativity by at least one extra dimension had a familiar ring. Many theorists of the time were at least somewhat acquainted with five-dimensional Kaluza-Klein theory through a section about that topic in a popular relativity textbook. The text was authored by Peter Bergmann, who worked with Einstein on several attempted five-dimensional unified field theories. By the time of Einstein's death in 1955, virtually no one was working on Kaluza-Klein type models—adding an extra dimension to ensure enough mathematical space to include electromagnetism along with gravitation. Part of the abandonment of those ideas was the lack of viable solutions. Another reason was that, since the time of Kaluza and Klein, the physics community had come to accept the idea of four fundamental forces—including the strong and weak nuclear interactions—not just two.

Essentially no one saw a twenty-six-dimensional theory of the strong force as a promising path forward, not even Lovelace. Yet a small group of theorists noted that higher-dimensional leap and would keep it in mind as new developments in unification progressed.

SUPERPOWERS

One limitation of hadronic string theory is that it modeled the strong force's agent as a completely different beast from anything in the particle zoo. A string or rubber band clearly isn't the same as a mathematical point. For parsimony's sake, a comprehensive theory of the subatomic world should have uniform ingredients. To be consistent, everything should be either point particles or vibrating strands. The latter would require reaching beyond just a stringy description of the strong force, and modeling all forces and particles as various kinds of strings.

By then, physicists realized that force carriers—the special kinds of particles, such as photons, exchanged to produce fundamental natural interactions, such as electromagnetism—were generally represented by bosons and matter constituents by fermions. Bosons are particles with integral spin quantum number: generally 0, 1, or 2. They tend to be sociable particles, grouping together into common quantum states until rising temperatures force them out. That is, at the lowest temperatures bosons often share the same ground state, but at higher temperatures they are distributed among higher energy states. Fermions, on the other hand, have half-integer spin: $½$, $³⁄₂$, etc. In contrast to bosons, fermions need breathing room and never willingly cluster into a shared quantum state. If two electrons, which are fermions, happen to share the lowest energy level of an atom, they must differ in their spin states. If one is spin-up, the other must be spin-down—like the opposite-oriented batteries in some flashlights.

Representing carriers of the strong force, hadronic strings fall resolutely into the boson category. To model the relevant quarks and antiquarks in similar fashion, researchers realized, would require fermion strings. In 1971, theorist Pierre Ramond of the University of Florida brilliantly derived a way of generating such half-integer spin vibrations through a method known as supersymmetry (SUSY).

Supersymmetry is a theoretical transformation that bumps up or down the spin of a subatomic constituent by half-integer spin increments, effectively turning a boson into a fermion, and the converse.

Like a bizarre psychology experiment applied to subatomic society, it converts crowd-seekers into wallflowers, and social pariahs into the life of the party. Bosonic strings that love to unite as force agents become fermionic strings that, as matter constituents, would, if freed from forces, simply wander off by themselves. Supersymmetry thereby strives to find a way to explain all things via a single initial ingredient that diversified in the distinct past into the stuff of energy and matter.

Within a few years after Nambu and others introduced the hadronic string proposal it was already on the way out. Instead, theorists developed a quantum field theory of quarks and gluons similar in many ways to the electroweak unification model. Rather than electric charge with two options, positive and negative, quantum chromodynamics (QCD) includes a "color charge" coming in three varieties: red, green, and blue. The opposites of those "colors" would describe antiquarks. Note that the term "color" in QCD has nothing to do with visible hues, but rather is a shorthand for describing something that is complete only as a blend of three, similar to mixing tinted light together to get white. Baryons always come in sets of three quarks, so the analogy made sense.

With electroweak theory and QCD shown to be renormalizable—a technical term meaning that all troubling, infinite terms could be canceled out, ensuring reasonable finite theories—gravity seemed the odd force out. No renormalizable, quantum theory of gravity existed, despite valiant efforts by Bergmann, DeWitt, and many others. Calculation after calculation produced frustratingly irremovable infinite terms. Fortunately, the rise of supersymmetry brought new hope to the struggle.

Renormalization might be envisioned by imagining a married couple that spends $3,000 each month on rent, utilities, groceries, and assorted other expenses. One member of the pair, a schoolteacher named Jane, earns all the income for both of them—which turns out to be $3,000 per month as well, after taxes are deducted. The other, Joan, is in charge of shopping and paying all the bills (as well as engaging in volunteer activities). To make a point during a discussion of finances about her contributions to the household, Jane emphasized her wages, ignored expenses, and pointed out the incredible amount of money they'd have if she kept working for decades. On another occasion, in a moment of anxiety, Joan brought up all their bills, overlooked Jane's steady income, and opined that in the coming years they'd be on the road to ruin. Indeed, if they somehow both lived forever (good genes, perhaps), their total income and total expenses would each add up to infinity. Luckily, a ledger sheet in which each month's losses are canceled with its gains to show a consistent balance would offer a modicum of realism. Similarly, even if a physical theory has terms that "blow up" to infinity, as long as, in the process of renormalization, they can be grouped in a way to cancel each other, it becomes acceptable for all practical purposes. Researchers hoped that supersymmetry would serve the function of helping provide the balance needed for the potential renormalization of a gravitational quantum field theory.

Developments in applying the concept of supersymmetry to a potential quantum theory of gravity proceeded rapidly. In 1973, Julius Wess and Bruno Zumino developed a means of applying supersymmetry to particles, rather than strings, in a quantum field theory. Then, in 1975, at a lecture delivered at Princeton, Caltech physicist John Schwarz announced a remarkable discovery. Within the context of supersymmetry, as applied to string theory, he and French physicist Joël Scherk found that bosons of spin 2 naturally emerged, and that they could be identified as the force carriers of gravitation in a quantum theory. Those preferring particle explanations quickly grabbed hold of the idea, and deemed the spin-2 exchange bosons connected with gravitation "gravitons."

Along with another French scientist, André Neveu, Schwarz explored the mathematics of superstrings: supersymmetry applied to strings. Like Lovelace, they found that extra dimensions were needed for a viable theory. However, rather than twenty-six dimensions in total, they found that ten would do the job. A ten-dimensional theory still seemed rather outrageous, but at least it was more palatable than twenty-six. Getting back down to four would be ideal, but the researchers determined that ten was the minimum number of dimensions that could possibly support vibrating strands that had boson and fermion modes.

If there are ten dimensions in the universe, and we perceive only four of them as ordinary space-time, why don't we observe the other six? That dilemma was cleverly addressed by French theorist Eugène Cremmer, working along with Scherk at the École Normale Supérieure in Paris. Cremmer and Scherk wrote several key papers introducing the notion they called "spontaneous compactification."

Image 17. Superstring theory co-founder John Schwarz, co-recipient of the 2014 Fundamental Physics Prize. Credit: AIP Emilio Segrè Visual Archives, Physics Today Collection.

The history of Kaluza-Klein theory, as we recall, included various attempts to explain the non-observability of the extra dimension—in that case, the fifth. In his five-dimensional unification model Kaluza artificially imposed a "cylinder condition" that eliminated all mathematical terms potentially allowing direct detection of the extra dimension. Its presence could be felt only indirectly. Klein imposed a rule that the extra dimension was circular and too minuscule to be seen, like a microscopic ring. Similarly, Einstein and Bergmann pictured a slender tube viewed from so far away that its thickness, representing the fifth dimension, was unnoticeable.

Cremmer and Scherk upgraded Kaluza-Klein theory with a novel way of explaining why the extra dimensions—six, in the case of superstrings—couldn't be observed. Employing the mechanism of spontaneous symmetry breaking, they constructed a ten-dimensional representation of physical fields that, at highest energy, would favor a state in which all of the dimensions are equal in scale, but, at somewhat lower energy, would drop into a state in which six of those dimensions would become tiny and compact. Like a flower wilting as temperatures dropped below freezing, for superstring theory, the six extra dimensions would shrivel. If the compactification was particularly simple, the cluster of extra dimensions would resemble a higher-dimensional generalization of a torus (doughnut-shape): the product of six tiny circles along distinct dimensions.

Within the ten-dimensional enclave, including four normal and six compact dimensions, each superstring would buzz with energetic vibrations. The nature of the vibrations would determine the equivalent particle properties such as mass, charge, and spin. As in manipulating the strings of a guitar to strike various chords, the theory would produce a wide range of features with a single instrument. Changing a feature called the "string tension" would affect the types of vibrations produced, and thereby the states of the particles represented—much like tightening a guitar string affects its sound. Everything would transpire on the minuscule Planck scale, at lengths of approximately 10^{-33} inches, so all experimenters would record at ordinary detector energies would be the conventional particle properties. That's so small that if an atom was enlarged to the size of a galaxy, the strings composing its electron, quark, and gluon constituents would be barely the size of fleas. Yet the notes plucked on those impossibly tiny cords would resound in all natural things. With their penchant for applying mathematics to music and nature, the ancient Pythagoreans would have marveled at the harmonious connections that superstring theory entails.

Yet, from the late 1970s to the early 1980s, the mathematical elegance of superstrings was lost on the mainstream theoretical physics community. Far more interesting to researchers was applying supersymmetry to a particle-based field theory of gravitation and other forces, called supergravity. Particle theories—with their splendid use of intact and broken symmetries—were so successful in the case of electroweak theory and QCD that it seemed foolish to abandon them. Besides, as it was impossible to test directly the notion of minute, oscillating strands on the Planck scale, there seemed little motivation to believe in them. The idea of point particles, in contrast, was a time-honored tradition in physics, dating at least as far back as Newton's "corpuscles." Particle-based SUSY, focused on incorporating spin-2 gravitons into a viable, renormalizable quantum field theory, seemed the way to go.

In 1976, Daniel Freedman, Sergio Ferrara, and Peter van Nieuwenhuizen, and independently Deser and Zumino, published the first supergravity theories. Supergravity refers to any supersymmetric quantum field theory that naturally encompasses spin-2 fields representing gravitational exchange bosons. The early versions of supergravity were housed in conventional four-dimensional space. That turned out to be their starter home, though. Theorists soon realized that higher dimensions would be needed to accommodate all the energetic fields (representing the various particles and forces) of the Standard Model, along with gravitation. Within a few years, expansion to an eleven-dimensional abode became the trend.

Among the pioneers of eleven-dimensional supergravity was French physicist Bernard Julia, starting with a seminal 1978 paper by Cremmer, Scherk, and him. The paper made good use of Cremmer and Scherk's spontaneous compactification technique to justify why the universe appears four-dimensional under all direct physical probes. Julia shared their enthusiasm for the method, and for the prospect of unity in general.

Several years earlier, Julia had learned about supersymmetry from Scherk and realized its potential to officiate a happy wedding between the bosons and fermions. A life-changing opportunity arose for Julia with a grant to do research at Princeton, which he had accepted with excitement. There, he attended Schwarz's 1975 lecture, which alerted him to how supersymmetry was key to quantizing gravitation. He also became close friends with brilliant young theorist Edward Witten.

Gravitational theory is part of Witten's heritage. His father Louis

Image 18. American theorist Edward Witten, recipient of numerous awards including the 1990 Fields Medal, who published key papers in superstring theory, M-theory, and related fields. Credit: AIP Emilio Segrè Visual Archives, Physics Today Collection.

Witten (now over 102 years old and still alive at the time of this writing) was involved in general relativity research and presented at the 1957 Chapel Hill conference. Growing up in diverse, working-class Baltimore, young Ed also became interested in the politics of social justice. Thus, it reportedly was a shock to his system, at first, when he ended up in the snooty, upper-crust environment of Princeton.

"When I first met Witten in Princeton, he felt isolated," Julia recalled. "I fed him a lot of cookies. He was very depressed."[2]

Motivated in part by the findings of Schwarz and Scherk, the two came to share an interest in higher-dimensional unification theories. Each reached out, at various points, to elder physicists to find out more about the history and setbacks of Kaluza-Klein theory. Feeling that the theory was ripe for a comeback, they wanted to learn from past experiences and finally make it work.

In 1979, as part of an international celebration of the hundredth anniversary of Einstein's birth, the second Marcel Grossmann Meeting on General Relativity took place in Trieste, Italy. Dirac was one of the esteemed guests. Attending the conference, Julia was excited to have the opportunity to speak with him. Julia asked him about the prospects of extending the Dirac equation (describing relativistic electrons and other fermions) into extra dimensions. In Julia's perception, Dirac seemed uncomfortable with the question, replying that he'd have a response, "maybe later."[3]

In 1981, Witten wrote to Bergmann, who was then a professor at Syracuse University, asking him about choices he made in his five-dimensional work with Einstein during the late 1930s. In particular, one assumption they made freezes the value of the gravitational constant, rather than allowing gravity to freely vary as a scalar field. If they had relaxed that assumption, the Einstein-Bergmann model would have been similar, in some ways, to Brans-Dicke theory (discussed in Chapter 4). Witten also shared with Bergmann some of his own recent Kaluza-Klein explorations. Bergmann responded very cordially, agreeing that his theory with Einstein had certain limitations, and welcoming Witten's new contributions.

One of Witten's important results from that period served to validate the hunch by Cremmer, Julia, and Scherk that supergravity would shine the brightest in eleven dimensions. He calculated that eleven would be the minimum number of dimensions needed to accommodate all the symmetry groups of the Standard Model, along with gravitation in a supersymmetric unified field theory. Thus, supergravity needed at least eleven dimensions to do its job.

Unfortunately, by the mid-1980s, supergravity's luster began to fade with the realization that none of its versions seemed completely renormalizable. In doing calculations with supergravity models, theorists found that they could begin the process of canceling infinite terms to get a finite solution. However, if they kept going to higher-order calculations involving virtual exchanges of fields (technically known as "three loop" contributions), infinite terms that they couldn't get rid of began to arise. Like the arcade game of Whac-A-Mole, they couldn't eliminate all the troublesome mathematical varmints.

SUPERSTRING REVOLUTIONS

By 1984, Schwarz was working closely with Michael Green of Queen Mary College, part of the University of London, continuing to pursue superstring theory, rather than supergravity. They saw a clear advantage. Unlike supergravity, superstring theory didn't include infinite terms that needed to be eradicated to make it finite. That's because strings are finite in size, rather than infinitesimal like point particles. Thus, whenever the reciprocal (multiplicative inverse) of length appeared in string theory formulas, the answer would be finite as well. Renormalizing the theory was not required, as it was finite all by itself.

At a conference in Aspen, Colorado, that year, Green and Schwarz announced another promising result. At least some forms of ten-dimensional superstring theory were wholly free of anomalies. Gravitational anomalies pop up in some versions of supergravity and other attempted quantized theories of gravitation as terms that contradict some of the main rules of general relativity. That's not acceptable for unified field theories, because we know that general relativity applies to a huge range of astrophysical and cosmological situations. Hence, Green and Schwarz's anomaly-free theory was met with widespread praise—especially after Witten replicated and confirmed their findings. Nobel laureates Gell-Mann and Weinberg offered their blessings and support. Jumping onto the superstring train, most mainstream high-energy theorists largely left supergravity behind—except for ways the two theories converged. The first superstring revolution had begun.

During the process of spontaneous compactification, the ten-dimensional superstring universe would split into two parts: a six-dimensional compact glob, and ordinary four-dimensional space with its swarms of vibrating strings. Some of these strings would be "open": having loose ends like floppy spaghetti. Those would represent most fermions and bosons. A major exception would be gravitons—represented by "closed" strings with connected ends, like tiny onion rings. For technical reasons, because a certain type of supersymmetry applied to the four-dimensional realm, its properties would be tied to the symmetries of the compact six-dimensional glob. Thus, to reproduce the Standard Model in ordinary space, the compact space would need to be arranged just right.

In 1985, Witten collaborated with Philip Candelas of the University of Texas at Austin and Gary Horowitz and Andrew Strominger of the University of California at Santa Barbara to examine the needed properties of the compact six-dimensional space. Generalizations of rings and doughnuts proved unsatisfactory in reproducing the Standard Model in conventional space. Rather, they explored gnarly higher-dimensional figures called Calabi-Yau spaces, named after mathematicians Eugenio Calabi of the University of Pennsylvania and Shing-Tung Yau of Harvard, for their respective works on twisted geometries. There are multitudes of such contorted configurations, each with a distinctive topology (such as how many holes it has) bearing on the physical properties of the space-time in question.

By the late 1980s, superstring theory had five distinct versions, and numerous choices for compactification. Having many options is not always a good thing, especially when trying to come up with a "theory of everything." It is like a bride bringing five hundred potential grooms to her wedding, and asking the priest to bless a perfect, everlasting matrimony with Mr. Right—whomever the priest thinks would work out best. Physics seeks a single perfect union, not a harem of topologies.

Joining the confusion was the introduction of membranes: quivering, two-or-more-dimensional shapes that joined vibrating, one-dimensional strings as the constituents of reality. While at first many string theorists resisted their introduction—thinking them an unnecessary complication—eventually they realized that membranes and superstrings had deep connections, called dualities. Dualities involve switching small amounts with large quantities, or the converse, for several different properties, such as thickness and interaction strength, and getting similar answers. Because membranes and strings could be transformed by certain mechanisms into one another, mathematically, they proved to be close cousins.

The full acceptance of membranes as part of the family arrived in 1995 when the "second superstring revolution"—as theorists soon called that period—began. Just as in the first revolution, Witten—with his brilliance, straightforwardness, and clout—played a pivotal role. At a conference held at the University of Southern California in February of that year, he delivered a compelling talk explaining how the five main versions of string theory could be united, along with supergravity, in a ten- or eleven-dimensional theory that included membranes along with strings. He dubbed the combination "M-theory." Unusually vague when explaining what the "M" referred to, joking that it could stand for "magic," "mystic," or "mother" (as in "mother-of-all-theories"), as well as "membrane," many researchers presumed that it simply stood for the latter. Perhaps if he had just said "membrane theory," many might have wondered whatever became of good old superstrings. Some of those working on membranes for years, such as physicist Michael Duff, considered the enigmatic designation a slight to their contributions. Lecturing at a 1997 meeting in Trieste, he said that Witten calling it "M-theory," rather than "membrane theory," was a "Pyrrhic victory."[4]

Despite presenting itself as a unifying mechanism, there are many ambiguous aspects to M-theory. For one thing, because of dualities, the total number of dimensions might be considered either ten or eleven. A collaboration between Witten and Czech physicist Petr Horava showed how both answers would be right under different circumstances. In an influential 1996 paper, Horava and Witten showed how superstrings living in ten dimensions would thicken along an extra dimension if their coupling constants (interaction strengths) were cranked up. (Technically that would happen because of a combination of dualities.) Miraculously, they'd turn from thin, vibrating strands in ten dimensions to pulsating two-dimensional membranes, or 2-branes, along an eleventh dimension. The ten-dimensional superstring realm would enlarge into an eleven-dimensional "brane world," as such a resulting state-of-affairs was deemed—"brane" being shorthand for "membrane."

Brane worlds have a complex arrangement indeed. After compactification, there are four space-time dimensions, three of space and one of time, accessible to open and closed strings alike. That is the material world we experience, with open strings representing the force bearers and matter components, and closed strings representing the gravitons conveying the gravitational interaction. As theorist Joseph Polchinski and others discovered, however, there would be a marked difference between open and closed string behaviors relative to ordinary 3-D space, known as a "Dirichlet brane," "D-brane," "3-brane," or simply "brane" for short. Like flies to flypaper, the open strings would remain attached to the brane of space, while the closed strings would be free to move off it. Gravitons, as closed strings, could flit from our brane to a greater region spanned by an extra spatial dimension, called the bulk. Allowing in only gravitons, but forbidding electrons, quarks, photons, and other familiar particles, the bulk would be an exclusive enclave for sure. Finally, to complete the picture, six more dimensions would be curled up into a variety of compact, Planck-scale Calabi-Yau spaces. The topology of the internal space in question would set the parameters of the physics in the other dimensions—in some cases offering something like the Standard Model, in others weird rules beyond our comprehension.

SURVIVAL OF THE FITTEST IN THE MULTIVERSE ARENA

In the mid-2000s, a series of papers by physicist Michael R. Douglas of Rutgers University, along with several other researchers, jolted the theoretical physics community by estimating the number of distinct Calabi-Yau spaces for various string and M-theory models to be enormous amounts. One paper, with then-Rutgers PhD student Sujay

Ashok, focused on a particular type of string theory and arrived at a whopping figure of 4×10^{21}: four billion times a trillion. That was followed by a more general overview of the topic in collaboration with Shamit Kachru of Stanford, concluding that the number of configurations must be at least 10^{500}.[5] Such a number is beyond astronomically large, as there is nothing in astronomy that matches it. Wrapping one's mind around the number 1 followed by 500 zeroes would be truly unfathomable. Yet, strange at it sounds, each of those possibilities would correspond to a different set of rules for the universe. If all of those are in any sense physically realizable, then reality would comprise a multiverse of unimaginable complexity. The number of universes in the multiverse would be far, far greater, in that case, than the number of atoms in the observable universe.

Contemplating such immensity might inspire some to flee in horror or laugh in amusement or derision. However, for some theorists it represented a challenge and perhaps an opportunity, pertaining to a key mystery in science: Why is the cosmological constant, as measured by astrophysicists, so small, but not exactly zero?

While the cosmological constant was first introduced by Einstein to stabilize the universe, and then discarded by him after Hubble's discovery that distant galaxies move away from us, suggesting an expanding universe, in recent years it has taken on new meanings. First, as Weinberg and many others have suggested, it is a way of characterizing the bedrock energy density of the quantum vacuum: the state as close to emptiness as possible, but brimming, nonetheless, with random quantum fluctuations. Because of the production and decay of virtual particles, like porpoises rising from the sea and then returning, the vacuum is never truly empty. For any quantum field theory, therefore, including the Standard Model, supergravity, various versions of superstrings, etc., the vacuum energy density might be calculated, and identified as an effective cosmological constant. It would help drive expansion of the universe in the way that a cosmological constant would do.

Second, because of the Nobel Prize–winning 1998 discovery that the expansion of the universe is speeding up—determined by supernova measurements conducted by two teams of researchers, one headed by Saul Perlmutter and the other by Brian Schmidt and Adam Riess—physicists believe that the universe has an invisible, outward-pushing ingredient dubbed dark energy. One of the simplest ways of modeling such dark energy is by adding back a cosmological constant to general relativity. That is equivalent to including an extra substance with "negative pressure": the tendency to push space outward.

Frustratingly, while standard quantum field theory predicts a humongous cosmological constant, the measured value of the accelerating growth of the universe implies a very small one. The mismatch is glaring, and wholly unexplained by conventional physics.

However, in the string theory multiverse, as characterized by the various Calabi-Yau configurations spawned by compactification, there are universe models with extremely low (or effectively zero) cosmological constants. That has led noted physicists such as Susskind and Weinberg to explore the idea that the observed universe might be a rare bird in the multiverse aviary. Ours might have an abnormally low cosmological constant. Yet, because such a low cosmological constant would lead to slower expansion in the early phases of the universe and allow more opportunity for galaxies, stars, planets, and living things to form, unless it was low we wouldn't be here to note that fact. Consequently, through anthropic reasoning, we can narrow 10^{500} possible configurations down to the ones with the lowest cosmological constant.

Susskind has had a lifetime sense of wonder. He enjoyed mathematics as a child, but kept that quiet from other kids in his tough neighborhood. As a young adult, he informed his father—a plumbing contractor who knew nothing about theoretical physics—"I want to do the kind of thing that Einstein did."[6]

Respected for his long-standing connection with string theory, having been one of the pioneers of the field, Susskind had started to realize by the turn of the twenty-first century that the once-straightforward idea was getting very messy. However, like an unlimited buffet with myriad dishes and platters, the more options, the greater the chance of finding something delectable. Fancy a savory, low-calorie "cosmological constant light," delicate enough to avoid a bloated universe? Perhaps a whopping smorgasbord of Calabi-Yau configurations would carry that item.

In 2000, an influential paper by Raphael Bousso of Stanford and Joseph Polchinski of University of California, Santa Barbara, appeared with a novel attempt to explain why the measured cosmological constant term propelling the accelerated expansion of the universe is so close to zero, but still nonzero. They showed how there are certain unusual modes of compactifying the extra dimensions in M-theory that would lead to energy field configurations producing bubble universes (in eternal inflation) that end up with small cosmological constants. However, even if such situations rarely happened, as long as they existed at all, the Anthropic Principle would single them out as being ideal for the eventual emergence of sentient life. Consequently, M-theory compactification would provide a narrow window for the favorable cosmological conditions that—in accordance with the Anthropic Principle—we know must have happened for us to be here.

As Bousso and Polchinski emphasized in their paper:

> The appearance of the anthropic principle, even in the weak form encountered here, is not entirely pleasant, but we would argue that it is necessary in any approach where the cosmological constant is a dynamical variable. That is, a small value for the present cosmological constant cannot be obtained by dynamical considerations alone.[7]

In 2003, with even more confidence about wielding the power of the Anthropic Principle, Susskind combined it with optimization theory, the branch of mathematics associated with techniques for minimizing or maximizing functions, to try to tame the cosmological-constant aberration. In his essay "The Anthropic Landscape of String Theory," he showed how a multiverse of enclaves with different string vacua (bedrock energy conditions for string theory based on various ways to compactify the extra dimensions into Calabi-Yau spaces) would likely contain some with a small but nonzero cosmological constant value. Characterizing the array of string vacua as a "landscape," he envisioned how the bubbles formed in eternal inflation could help explore that domain by sampling each of the possibilities. As he wrote, "The incredible smallness and apparent fine tuning of the cosmological constant makes it absurdly improbable to find a vacuum in the observed range unless there are an enormous number of solutions with almost every possible value of [the cosmological constant]."[8]

The string landscape, in Susskind's view, could allow for a game of survival of the fittest—with the measure of fitness being a low but nonzero cosmological constant leading to a relatively slow-growing universe suitable for planetary system formation and ultimately life. Eternal inflation would provide the contestants in that epic cosmic battle. Each cosmic bubble, produced in accordance with Linde's theory, would have a cosmological constant, determined by the topology (number of holes, twists, and so forth) of its compactified Calabi-Yau space. As the bubbles continued to expand upon the close of their respective inflationary bursts, almost all of them would have high cosmological constants and enact rapid-fire further acceleration to oblivion. They'd be the losers, from the perspective of us living beings. Only a precious few, having tiny but nonzero cosmological constants, would pace their growth, allow gravitation to forge stable structures, build up planetary systems with worlds suitable for life, and have beings on it marvel about how special the universe truly is.

The message that the Anthropic Principle and multiverse would both be needed to justify the small, finite value of the cosmological constant was greatly amplified when Weinberg lent his endorsement. Like Witten and Susskind, Weinberg was a child prodigy. Born in 1933 in New York City to immigrant parents, by age sixteen he was enraptured with

Image 19. Nobel laureate Steven Weinberg, one of the key developers of the Standard Model of particle physics, and a leading advocate of multiverse ideas. Credit: AIP Emilio Segrè Visual Archives.

learning more about the physical world. After receiving an undergraduate degree from Cornell in 1954, he headed to Bohr's Institute for Theoretical Physics in Copenhagen for a visiting research position at the age of only twenty-one. Returning to the US, he completed a PhD from Princeton in 1957.

The 1960s and early 1970s were a highly productive time for Weinberg. Working at Berkeley, MIT, and Harvard, his extraordinary contributions to electroweak unification earned him not just the Nobel Prize but also the utmost respect from the physics community for his meticulous approach to quantum field theory. Meanwhile, in the public mind, his popularizations—on the drier side, with little hyperbole, but rather containing considerable useful information explained carefully and clearly—elevated him to the role of a trusted authority in science.

Therefore, many physicists were surprised when, in 2005, Weinberg, by then a professor at the University of Texas at Austin, announced his support for multiverse ideas and for using the Strong Anthropic Principle as a way of sifting through the alternatives. Some were enthralled; others disappointed. In the latter category was Peter Woit of Columbia, who wrote in his blog:

> What Weinberg sees as "excitement" is what some others have characterized as "depression and desperation." His "radical change in what we accept as a legitimate foundation for a physical

theory" seems to be to give up on the idea of a fundamental theory that predicts things and instead adopt the "anthropic reasoning" paradigm of how to do physics.[9]

Weinberg's paper "Living in the Multiverse" was delivered at Trinity College, Cambridge, presented on the physics "arXiv" (pronounced "archive" because the *X* is meant to represent the Greek uppercase letter "Chi"), and later published in a collection edited by Bernard Carr, *Universe or Multiverse?* In it, after briefly surveying the history of unification efforts to date, he argued that the next step forward would likely involve using the sieve of our existence, rather than standard calculations, to fine-tune the universe. Anthropic considerations would be necessary to filter out all but livable branches of a complex multiverse, associated with an immensely complicated string landscape. He called that step "a new turning point, a radical change in what we accept as a legitimate foundation for a physical theory."[10]

Weinberg later described how he came to advocate for applying the Anthropic Principle to the string multiverse:

> I've been bothered by the cosmological constant for a long time, wondering why it wasn't huge. . . . We knew it was many, many, many orders of magnitude less than you would guess from a back of the envelope calculation. That was a great mystery. In my lectures at Harvard, I went through all the proposals that had been made that I knew about. I had negative things to say about them all, except for the possibility of an anthropic explanation. That is that we're in only one sub-universe of a multiverse and in most of them the cosmological constant is very large. And it's only in the ones where it's small that life could arise because it's only in those that galaxies and stars could form.[11]

Susskind and Weinberg were hardly the only physicists of their time concluding that anthropic reasoning and some kind of multiverse could very well be required to move forward. Some of the others who published on that topic in the late 1990s and early-to-mid 2000s included Swedish-American physicist Max Tegmark, Israeli-American astrophysicist Mario Livio, and English cosmologist Martin Rees. Theorist Juan Maldecena of the Institute for Advanced Study nicely summed up the reasons for such advocacy:

> I think that the multiverse explanation is the simplest explanation of the cosmological constant in a theory where it is not a fundamental parameter. Suppose we have a theory where the cosmological constant is in principle calculable, as in string theory, where it depends on the details of the internal geometry. Then its natural value is very large in magnitude, but it could have either sign. As we vary the details of the internal space it would vary in an apparently random way. So, if you have a huge number of solutions, it is possible that one of them has the observed small value. . . . A good question is how we should think about this multiverse. Whether they all exist at the same time in distant regions, or that they just all exist as mere possibilities, as in the many worlds of Everett.[12]

Physicist Gordon Kane has pointed out that eternal inflation requires certain initial conditions. Therefore, the absence of those prerequisites in a wide variety of spaces may weed out much of the string landscape before the bubbles form. That would narrow down the possibilities considerably, as the vast majority would not be fated to become actual universes. As he commented, "Inflation doesn't occur in the world automatically. . . . And any theory that doesn't inflate doesn't matter. It may be ninety-nine percent of all multiverse worlds don't inflate."[13]

In other words, it may be the case that of the 10^{500} modes of compactification, the overwhelming majority produce spaces that are absolute duds. If only a tiny percentage of modes yield quantum field theories that create scalar field energy profiles suitable for triggering inflation, then the problem of narrowing down would be less daunting. Finally if, of that tiny portion, only a minute fraction engender small, but nonzero, cosmological constants, it might conceivably be the case that string theory's embarrassment of possibilities would narrow significantly. String theorists' dream of crafting a theory of everything would then be far more viable.

KEEPING IT SIMPLE

Suppose a young child is asked by her teacher to design a spaceship out of cardboard, crayons, and colorful paper. Putting her imagination to good use, and working at amazing speed, she builds a model with dozens of rooms, including a bakery to make cookies, a garage for space-tricycles, a swimming pool, a time machine room so that they can travel back in time to fix any mistakes, a pizza parlor, and a castle on top, in case they want to host any alien royalty. She gets five stars for creativity.

Later in life, as a very intelligent young woman, she becomes an aeronautics engineer designing real vehicles. By then, she learns that economy, sleekness, and efficiency are key. Ergonomically, the aircraft and spacecraft she helps create are planned to be the simplest possible that remain safe, effective, and test drive very well in general. She gets a promotion for efficiency and reliability.

Scientists, in general, seek simple, testable truths that match previous experimental evidence in a straightforward way while offering clear predictions for future investigations. Like efficient vehicles rather than pleasure craft, scientific models are meant to avoid extravagant, unnecessary elements and to judge success by means of repeatable tests with measurable outcomes. Speculation offers fun science fiction but not always usable science.

Where do superstrings and M-theory fall? That is the multibillion-dollar (or, more accurately, euro) question that has become one of the objectives of the Large Hadron Collider, the giant particle-smashing ring operating deep beneath the Swiss-French border. In collision debris, researchers have been seeking evidence of supersymmetric companion particles (fermion counterparts of bosons, and the converse) and of echoes of the presence of higher dimensions. No such proof has been found so far, in many years of operation. Yet scientists who believe that those theories are simple and elegant have not given up hope. Even if such evidence eventually arrives, it would be nigh impossible to imagine directly testing the hypothesis of a string landscape, given that the twisty topology would act on the Planck scale, corresponding to ultra-high energies well above scientists' ability to generate. Nonetheless string and M-theory have continued to attract a bevy of researchers convinced that they are the way forward in unifying reality.

Meanwhile, dissenters continue to question whether theoretical physics has become a matter of faith. Leading lights such as Feynman, Wheeler, and Glashow never liked string theory. Steinhardt has found in the string theory multiverse many of the same issues as eternal inflation. He has found it similarly not predictive—offering so many options that it couldn't readily be narrowed. As he opined:

> String theory [seems to have] a similar problem as inflation. It doesn't produce a single type of particle physics. It produces a nearly infinite variety, like the multiverse—an exponential number—of different possibilities called a landscape of possibilities. And if that landscape picture is correct, it actually has the same sort of problem that inflation does, which is that it gives too many possible outcomes. And so, like inflation, it's not really telling us why things are one way versus another way.[14]

Based on his judgment that multiverses of all types—including eternal inflation and the string landscape—don't offer predictable science, Steinhardt has joined with other theorists in creating and advocating for cyclic universe models, also known as bouncing universes. Their model is not to be confused with the Conformal Cyclic Cosmology developed by Penrose—a way of using mathematical transformations to join together a bleak, empty ending of one cycle of the cosmos with the fiery beginning of the next, engendering an ouroboric closed loop.

Curiously, while eschewing the string landscape, the models of Steinhardt and his collaborators are based squarely on a consequence of M-theory. That is, they are brane worlds—which ordinarily include Calabi-Yau compactification of their six extra dimensions, as well as maintaining one large extra dimension. Seemingly, that compactification would create a string landscape, and hence a multiverse. Therefore, ironically, while their proposal aims to avoid a multiverse, it might not be able to do so.

Cyclic cosmologies, however, arguably hold a distinct advantage over constructs linked to string theory. They fully embrace time's arrow of nondecreasing entropy in each cycle. Moreover, they do so in a dynamic way, without needing to invoke the Anthropic Principle. Therefore, they speak to our intuitive sense that time passes unidirectionally—making exceptions only for periodic, universal resets. Addressing the reality of time, while also allowing for endless cosmic renewal, makes them attractive options.

I would not swear that some form of multiverse will remain forever unobservable, undetectable, even if Penrose's blisters on the CMB are not significant. If another bubble really hits and destroys us, the last scream will be "you were right!"

—Astrophysicist Virginia Trimble,
personal communication to the author

CHAPTER SEVEN

SEASONS OF REBIRTH

The Vying Cyclic Cosmologies of Paul Steinhardt and Roger Penrose

Cycles entice us, offering continuity instead of terminality. Nature's propensity for renewal stands in assuring contrast to the finality of death. From the tides, to the seasons, to the periodic patterns of the planets, the world seems to march to endless rhythms. Flowers pop up each spring as birds return from seasonal migration—offering colorful and tuneful tributes to natural resilience. No wonder many ancient religions embraced the twin notions of personal and worldly reincarnation.

In scientific cosmology, for many decades, the Big Crunch, and its oscillatory universe extension into a new Big Bang and a fresh cycle of time, were the go-to model for those who longed for an ever-renewing cosmos. Richard Tolman's finding, that oscillatory models build up entropy (a measure of the amount of disorderly waste energy) with each cycle, seemed to cast doubt on the notion of eternal oscillations. However, in the 1980s some physicists—most prominently, Stephen Hawking, for a brief period—started to wonder if the arrow of non-decreasing entropy (and forward time) might be linked to the arrow of spatial expansion. In that case, entropy would build up during the expansion phase following the Big Bang, then decrease during the contraction phase leading to the Big Crunch. Theoretically, in that case, eternal renewal would be possible. Note that Hawking changed his mind about such a reversal, after his PhD student Raymond Laflamme convinced him that he had made a mistake in his calculations. Now few physicists believe in such a turnaround in the direction of entropy accumulation.

Because the standard Big Crunch model is based on a closed, positively curved (like the surface of a sphere) spatial geometry and the slowing down of expansion over time, the 1998 discovery of the accelerating expansion of the universe, followed by the WMAP and Planck CMBR probes both indicating a flat spatial geometry, has nearly sent it to the scrap heap of history. It remains a faded alternative, lying in the wings, in case a considerable body of data somehow turns out to be wrong.

Meanwhile, those passionate about cosmic recycling have turned to newer models that don't include a Big Crunch. The Ekpyrotic Universe, proposed by Justin Khoury, Burt Ovrut, Paul Steinhardt, and Neil Turok, followed by the Cyclic Universe, proposed by Steinhardt and Turok, are each higher-dimensional models, based on the notion of periodically colliding branes in a brane world. The Conformal Cyclic Cosmology, proposed by Roger Penrose, involves conventional general relativity in four space-time dimensions, but applies a kind of mathematical trick, called conformal invariance, to allow the end of the current version of the universe to offer the seeds for a new cycle. Yet another model, loop quantum cosmology—developed by Martin Bojowald, Abhay Ashtekar, and others—posits that the early universe existed in a kind of frothy quantum foam left over from a previous period of contraction. Most recently Latham Boyle, Kieran Finn, and Turok have envisioned the Big Bang as a juncture between our universe, moving forward in time, and an "antiuniverse," moving backward in time. Such newer cyclic cosmologies have been dubbed "Big Bounce" universes.

THE CREATURE WITH TWO BRANES

Before being adopted by cyclic cosmology, brane worlds came into prominence in the late 1990s as a potential solution to a burning question: Why is gravitation so much weaker than the other forces? Known as the "hierarchy problem," the discrepancy between the relative strengths of the other natural interactions—electromagnetism, the weak force, and especially the strong force—and gravity's feeble pull is glaring. Why should nature have such a stark preference?

If you think gravitation is incredibly strong, picture this scenario. You run a strength contest using iron thumbtacks between gravity and magnetism. Gravity offers the heftiest contender in this immediate region of space, the whole Earth. Magnetism's designated contestant is a tiny bar magnet from a child's science kit. The thumbtack is placed on the ground with the magnet right above it. It immediately rises toward the magnet, leaving the ground behind. A winner is called—magnetism. The gravitation of all of Earth is no match for the force of a simple bar magnet. The lesson is that gravitation is effective mainly on astronomical scales, where other interactions play minimal or no roles.

In 1998, physicists Nima Arkani-Hamed, Savas Dimopoulos, and Gia Dvali proposed an innovative attempt to solve the hierarchy problem using a brane world model. In their much-cited paper "The Hierarchy Problem and New Dimensions at a Millimeter," they made use of the M-theory distinction between gravitons and other particles to explain gravitation's relative weakness. Their proposal became known alternatively as the "large extra dimension" or "ADD" model, named after the first letters of their surnames.

According to the large extra dimension idea, our three-dimensional space and another three-dimensional space, each serving as a Dirichlet brane, or brane for short, are separated from each other by approximately one millimeter (about 4 percent of one inch) along a higher dimension. That is, there are twin branes, one of them we live in and the other, our extra-dimensional neighbor across the way. The filling of that "brane sandwich" is a thin "bulk" through which gravitons, represented by closed strings, might pass, while all Standard Model particles, being open strings, are trapped on our brane.

Within our brane, therefore the Standard Model would reign supreme. It would be the main cast, with gravitation just the special guest star. Hence, unification would occur at the energy scale needed for restoring symmetry to the Standard Model, achievable in our colliders (such as when the Higgs boson was later identified at the Large Hadron Collider in 2012). To find unity, we wouldn't need to probe the Planck scale or reach the correspondingly much-higher energy associated with that regime. Gravitons would spend most of their lives in the bulk, and make contact with our brane only loosely. Such dilution of gravity's power into the region between the branes would thereby explain why that force is so much weaker than the others.

The ADD team explained that one key advantage of their model, over Planck-scale unification proposals, was that their model was testable. They predicted that at very short ranges (much less than one millimeter), gravitation would start to deviate from its usual behavior. Unfortunately, despite more than two decades of high-precision experiments looking for such short-range deviations, no discrepancies have been found so far.

The year after the ADD group introduced their model, physicists Lisa Randall and Raman Sundrum proposed their own methods of resolving the hierarchy problem—and also the cosmological constant problem—using brane worlds. They developed two different models—one with a single brane, the other with two. The single-brane model was innovative in that it included a warped geometry of the extra-dimensional space, and saddled it with a negative cosmological constant. Technically, such geometries are known as "anti–de Sitter" spaces.

Sundrum recalled how he became fascinated by the idea of applying brane world models to some of the thorniest problems in physics:

My interest lay in a problem known as the cosmological constant problem. The universe is expanding at a certain rate. It looks like we've expanded far, far slower [than predicted]. It naively looks like a problem with a very, very large cosmological constant. . . . This [interest] was even preceding the actual findings of acceleration. But . . . whatever the acceleration was it was very small, whereas theory preferred very big. And so there was already a puzzle. So I already had reason enough to be interested. I think that at least a lot of reasoning I'd been pursuing got a lot more interesting when we actually saw that there was not just a bound, but actually some finite acceleration.[1]

Rather than applying the Anthropic Principle to a string landscape—as would be suggested by others such as Susskind and Weinberg soon thereafter—the one-brane Randall-Sundrum model attempts to resolve the cosmological constant problem by adding a large positive value to a slightly smaller negative value to produce a tiny positive value for the overall cosmological constant. The large positive value is the vacuum energy of our brane, due to the baseline of quantum interactions in the Standard Model. The slightly smaller negative value is the negative cosmological constant of the higher-dimensional anti–de Sitter space in which our brane resides, surrounded by a bulk. Adding those together and producing a small positive cosmological constant explains why the accelerated expansion of the observable universe did not start to kick in until late in the game, permitting stars, planets, galaxies, and such ample time to form.

As in the ADD model, the Randall-Sundrum model also purports to resolve the hierarchy problem of gravitational weakness by envisioning gravitons being diluted into a bulk beyond our brane. Rather than trapping them in that bulk via another brane, however, the warped geometry does the trick. Like water trapped in a ditch alongside a road, the negatively curved bulk effectively confines gravitons to be close enough to our brane that gravity is weak but still effective. That means the extra dimension could be indefinitely large, eliminating the need to compactify it.

Tests were soon proposed to look for gravitons leaking into the extra-dimensional bulk. Alas, despite years of Large Hadron Collider experiments, no evidence of such leakage—in the form of missing energy or other measures—has been found to date. Still, many theorists continue to be hopeful in light of recent upgrades to that particle smasher.

THE PHOENIX AND THE FLAMES: THE EKPYROTIC AND CYCLIC UNIVERSES

None of the original brane world developers have suggested applying their ideas to cosmology, save trying to explain the smallness of the cosmological constant. They have preferred to concentrate on developing their models as steps toward unifying gravitation with the other forces, while avoiding having to deal with untestable Planck-scale physics. In essence, by means of an extra dimension inaccessible to ordinary matter and energy, they aspire to craft experimentally verifiable solutions to long-standing conundrums.

Steinhardt, on the other hand, arrived at brane worlds from a different angle. He was looking for alternatives to eternal inflation that would still reproduce known cosmological measures and lead to a flat universe with a scale-invariant Cosmic Microwave Background Radiation. Those considerations would lead him to co-develop the first brane world cosmologies.

Leaving one's legacy theory behind and venturing into largely uncharted territory is not a choice most physicists would have made. The inflationary universe, with which Steinhardt was closely associated, continues to draw much attention in cosmology—with many of its other founders, such as Guth, Starobinsky, and Linde, continuing to gather awards. Why abandon a successful venture with which one is deeply connected? Yet, as the son of an Air Force officer, Steinhardt

Image 20. Cosmologist Paul Steinhardt (right) who developed versions of the inflationary universe model in the 1980s and 1990s, and later co-proposed the Ekpyrotic and Cyclic Universe models as alternatives to inflation, photographed along with physicist Alan Goldman (left) and future Nobel laureate chemist Daniel Shechtman (center) at a 1987 meeting of the American Physical Society. Credit: AIP Emilio Segrè Visual Archives.

moved from place to place and base to base as a child before his family settled in Miami. Flexibility is in his upbringing.

Steinhardt has had a passionate interest in astronomy since childhood. During his teenage years, he owned a telescope and loved heading out, on steamy Florida nights, and gazing at the heavenly lights. That would require, as he recalled, ample protection against the persistent mosquitoes—especially when he headed to the Everglades to take advantage of its dark nocturnal skies.[2]

As an adult, following a position at the University of Pennsylvania, where he worked in the field of quasicrystals, as well as in co-proposing new inflation, he moved on to a professorship at Princeton. That's where, at the turn of the millennium, he began to ponder a new direction in his theoretical cosmology work.

Steinhardt recalled his path toward bringing M-theory into the study of the universe:

> The goal was to look for an alternative to the standard model—the inflationary/Big Bang picture. . . . When you get down only to a single competitor it's not always a healthy situation. It's much better to have two or more competing models, forcing you to think more carefully about your theories, your predictions and the observations. So that was really the motivation to try to look for something different. Now what had happened in the previous five years was string theory had introduced a lot of interesting new ideas into the game, thinking about fundamental physics on a microscopic scale—the idea of branes, the idea of extra dimensions. . . . Then I asked the question, could you do something interesting with it that would be a different kind of cosmology?[3]

Opportunity knocked when Steinhardt was attending a conference at Cambridge that included Turok, a fellow cosmologist, and Ovrut, a string theorist—a brainy trio that would soon go totally "brany" as well. Ovrut gave a talk about how branes merge in M-theory, detailing the implications for particle physics of such mergers. Following his lecture, Steinhardt and Turok pressed him on the cosmological implications of such brane collisions. During a shared train ride from Cambridge to London, the three physicists continued their dialogue—which Khoury, Steinhardt's student at the time, jokingly called a "brane storming" session—and began to sketch out a cosmology based on brane collisions.[4]

The four settled on a rather esoteric name for their idea, based on *ekpyrosis*, the ancient Greek word for the fiery end to a cosmic cycle, heralding the start of the next one. The Ekpyrotic Universe would draw its fire from a catastrophic collision between our brane and a neighboring one. In a scenario frightening enough to provoke nightmares among the anxiety-prone, each moment of our lives that neighboring brane would lie in wait, only a small distance away from us along an extra dimension. Its influence over us would be omnipresent in the form of what is called dark energy. At times in the past, and perhaps the future, it would crash into our brane, wiping everything out, and rendering space smooth and homogenous. The tremendous energy released by the collision would transform into the familiar repertoire of elementary particles. Timing variations in how the branes collided in different regions would produce the seeds of the minute temperature fluctuations found in the CMBR. The situation would be akin to a car bumper becoming unevenly dented due to a slow crash with another vehicle that impacted various parts of its surface at slightly different times. In short, the researchers constructed a model designed to reproduce the benefits of inflation, such as solving the flatness and horizon problems via a smoothing-out during collisions, and producing a similar CMBR profile due to timing effects, but without its baggage of ever-reproducing bubbles leading to an endless multiverse.

Steinhardt and Turok followed up their first proposal with an enhanced model, called the Cyclic Universe, which explicitly included periodic crashes between the branes, throughout eternity. Those catastrophic rendezvous would happen about once every trillion years, far longer than the current age since the Big Bang. Unlike the traditional Big Bang theory, the Cyclic Universe would have no start and no end.

As Steinhardt noted, "It's an advantage to have no beginning of time, because I think it's kind of a disturbing idea to go from no time to time."[5]

Right before each collision, the universe (on our brane) would contract and smooth out. Nicely (from Steinhardt's current perspective), that contracting phase would ensure that inflationary bubbles would never have a chance to form. Without a bubble multiverse, anthropic reasoning would not be needed to sort through the alternatives. There would be one and only one universe (aside from the other brane).

Steinhardt elucidated how inflation, and hence an inflationary multiverse, could be avoided:

I realized that one thing that happens during a contracting universe—which you would have before a bounce—is that you can't produce a multiverse. And the reason is, you're not ever inflating. The reason why you get a multiverse, is because you have rare quantum fluctuations which keep inflation going, which means gravity stretches them by a huge volume, so that they soon occupy most of the volume of the universe. And then quantum fluctuations on top of fluctuations keep repeating that process.

In a contracting universe, you can have rare fluctuations that keep you contracting, but you're contracting—so those regions aren't going to ever become important relative to regions that bounce and start expanding. So you immediately solve the multiverse problem.[6]

In February 2017, Steinhardt, along with physicists Anna Ijjas and Abraham "Avi" Loeb, published a critique of eternal inflation in *Scientific American*, offering Big Bounce–type theories, such as the Cyclic Universe, as an alternative. They argued that eternal inflation, by permitting an unlimited array of universes that could have an extraordinarily wide range of parameters, yet are far beyond observability, was essentially not falsifiable.

Dissent is a healthy part of science. Being doctrinaire is not. With growing triumphalism, many advocates of inflationary universe theory and string theory have posited those as the only games in town. In a way, they have positioned themselves in a manner similar to Einstein's clinging to local realism and determinism and Bohr's unwavering advocacy of the quantum role of conscious observation. While science might ultimately experimentally confirm proposals for eternal inflation and string theory unification, perhaps via indirect methods, that time likely remains a long way off. In the meanwhile, Steinhardt, Ijjas, and Loeb, recognizing the importance of principled criticism, felt obliged to speak up. By not making testable claims, they argued, eternal inflation had moved outside of the long-standing tradition of the scientific method.

Supporters of inflation took that as fighting words—deeming "unscientific" their cherished theory that so nicely reproduced CMBR results and explained so many gaps in the original Big Bang. In the same magazine, several months later, a group of physicists including Hawking and Linde issued a sharp rebuttal, writing:

> By claiming that inflationary cosmology lies outside the scientific method, IS&L [the authors of the earlier article] are dismissing the research of not only all the authors of this letter but also that of a substantial contingent of the scientific community.[7]

Is falsifiability absolutely necessary for a theory? Khoury, now a professor of physics at the University of Pennsylvania, doesn't think that rule is absolute. More than two decades since the Ekpyrotic Universe proposal, he remains open-minded about alternatives. While still fascinated by Big Bounce ideas, he recognizes the strength of inflationary models in matching the CMBR. As he noted:

> To me, this idea of falsifiability is, at some level, perhaps naïve. Ultimately the reason we give credence to certain theories over others is because they make predictions.... It's a pragmatic question. If eternal inflation—even if I cannot measure these other universes directly—makes predictions, and we test those predictions, and they turn out to be correct, then that would give me confidence in that idea.
>
> If you were to take a census of cosmologists right now, the vast majority would say they believe in slow-roll inflation. And why is this? Because the simplest inflationary models predict scale-invariant density perturbations.

Suppose we, in 10 years or so, measure primordial gravitational waves consistent with slow-roll inflation. Right then, I think that would, in many people's minds, close the deal for slow-roll inflation. . . .

The prospect of ever building a collider that would produce the inflaton particle on-shell is basically nil. We would never be able to directly test it. . . . So, I think it's a bit silly to apply that standard to the multiverse.[8]

Physicist Bernard Carr concurs that some fully legitimate theories might not yet have directly testable predictions. Regarding the multiverse, he "takes the view that it's on the border of physics and philosophy."

"I call it meta-cosmology," he notes, "in the sense that we don't have observational evidence for it yet, but it arises out of legitimate cosmological speculation."[9]

While they might disagree on the efficacy of multiverse models in resolving cosmological conundrums, Steinhardt, Khoury, and Carr have each emphasized open-mindedness toward competing theories. Neither eternal inflation nor bouncing cosmologies are at this point settled physics by any means. Each makes claims that are far from falsifiable as of yet, such as neighboring bubbles and proximate branes. Until conclusive evidence somehow points to the unmistakable victory of one of the competitors, respectful discourse is the way to go.

SEEDS OF REBIRTH IN THE COSMIC LOAM

The notion of being open-minded to alternatives similarly sits very well with Penrose, who, in his youth, admired how his mentor in physics, Dennis Sciama, switched from the steady-state theory to the Big Bang theory once evidence mounted for the latter. "Dennis was absolutely direct and honest, as a scientist should be, and I had tremendous respect for that," he remarked.[10]

Interestingly enough, Penrose didn't start out as a physicist. Rather, he was trained in mathematics. Born in Colchester, England, on August 8, 1931, to a doctor, Margaret Leathes, and a medical geneticist, Lionel Penrose, young Roger was certainly exposed to science as a youth. Yet, when it came time to pursue a career, Roger followed his older brother Oliver into the realm of pure math. After an undergraduate degree in that field from University College London, Penrose enrolled in a PhD program at Cambridge. That's where he met Sciama, who encouraged him to explore the nuances of general relativity. His subsequent development of black hole singularity theorems, and other research in theoretical astrophysics and cosmology, led to his 2020 Nobel Prize in Physics.

In recent years, Penrose has become an advocate for a distinctive model of reality, Conformal Cyclic Cosmology. It updates the Weyl Curvature Hypothesis (WCH) by means of using conformal invariance—an equivalence of geometries on different scales as long as they preserve shapes—to identify the close of our "aeon" (as Penrose calls each cycle) with the start of the next one. Recall from Chapter 4 that the WCH forces the entropy of the Big Bang to start out exactly zero.

Image 21. British mathematician and theoretical physicist Roger Penrose, developer of key theorems in general relativity, co-recipient of the 2020 Nobel Prize in Physics, and proposer of Conformal Cyclic Cosmology. Credit: AIP Emilio Segrè Visual Archives, Physics Today Collection.

One of Penrose's former students, Oxford mathematician Paul Tod, offered a key insight that led to his proposal. Weyl curvature offers a measure of the twistiness of space-time. If it is identically zero, to match an embryonic state of zero entropy, then space must start off expanding in a purely isotropic fashion. Lines tracking its expansion over time would flow as straight as arrows, something like the spokes of a wheel. Tod noted that, under conformal invariance, such a state would be fully equivalent to a large empty, expanding universe experiencing its bleak demise. That is, because conformal invariance preserves shape, but not necessarily size, small and large isotropically expanding spaces would be equivalent.

Penrose took up Tod's suggestion, and began to think about the endgame of the cosmos. Imagine an era, billions of years in the future, in which the sun, and stars similar in mass, have turned into petite white dwarfs. More massive stars have exploded in supernova bursts, and become either neutron stars or black holes. Flash forward billions more years, and all remaining stars, including white dwarfs as well as slow-burning red dwarfs, would sink into thermal equilibrium with the spatial vacuum. The universe would become a cold, desolate place, full of lifeless shards of formerly shining stars, as well as black holes. However, because of slowly emitted Hawking radiation, the black holes would gradually lose their mass. Cosmic conditions would become emptier and emptier, especially given the continued expansion of space. Even further into the future, Penrose envisioned, all known elementary particles would gradually disintegrate into cold radiation. Grand unification theories (GUTs) predict that protons and many other apparently stable elementary particles would all eventually decay. Ultimately, therefore, space, for all intents and purposes, would be empty—save its frigid background radiation.

In that state of utter oblivion, bereft of particles and any sense of change, the universe would suffer something akin to an identity crisis. It would suddenly "forget" its size, through a mapping of large to small in the process of conformal invariance. Instead of a big, old empty universe, it would become a small, vital young universe, expanding uniformly. With the Weyl tensor reset to zero, orderly energy would be abundant, heralding a flood of new particles. The universal clock would reset, and a new cosmic aeon would begin.

In a manner similar to Steinhardt and Turok's cyclic proposal,

Penrose's model wholly avoids the need for an inflationary era, after the Big Bang (in this case, a transition), to flatten and smooth out the universe. Rather, the flattening and smoothing transpires in the twilight age of the prior aeon, because of the accelerated stretching-out of space.

While mathematical models of the universe without reasonable observational predictions are a dime a dozen—the pages of theoretical journals are full of such papers—having a clear, experimental litmus test sets a theory apart. Suitably, in 2010 Armenian physicist Vahe G. Gurzadyan, in collaboration with Penrose, offered a very specific prediction about Conformal Cyclic Cosmology. They predicted that collisions of supermassive black holes in the previous aeon would make an imprint in the thermal radiation during the transition to our own aeon. Supermassive black holes are gargantuan bodies, millions of times more massive than ordinary black holes (of stellar origin). They lie in the centers of galaxies, such as the Milky Way. The result of such crashes, they determined, would be a bullseye-like signature of concentric circles of abnormally low temperature fluctuations in the CMBR. In other words, the perturbations from the average would be smaller than expected in patterns that looked like rings within rings.

Applying their analysis to the WMAP seven-year results, Gurzadyan and Penrose asserted in their paper that they had found several examples of such concentric circles in the CMBR. The claim made headlines in news stories. A story about the rings in the BBC News on November 27, 2010, proclaimed, "Cosmos may show echoes of events before Big Bang."[11]

Soon thereafter, several research teams applied their own statistical analyses to the concentric circles alleged to be signs of an earlier aeon. Each disputed the finding that the rings were anything unusual. On the contrary, their patterns fell squarely within the expected range of fluctuations in standard Big Bang cosmology, with dark matter, dark energy, and an early inflationary era. That is, nothing pointed to any novel phenomena verifying the predictions of Conformal Cyclic Cosmology.

Undaunted by the disputed results, in 2018 Penrose joined with another collaboration, including Daniel An, Krzysztof Meissner, and Pawel Nurowski, to claim another emblem in the CMBR sky of a previous aeon. Their analysis identified purported examples in the Planck satellite data of "Hawking points": blemishes caused by Hawking radiation slowly emitted in the last aeon from evaporated supermassive black holes. However, that paper was rebutted too. Statistical analysis conducted by Canadian astrophysicists Dylan Jow and Douglas Scott, published in 2020, found no evidence of Hawking points.

Another prominent group of opponents of string theory and the inflationary universe (at least in its standard description) who have crafted their own type of Big Bounce model are the advocates of loop quantum cosmology, an offshoot of loop quantum gravity. Developed by physicists Abhay Ashtekar, Lee Smolin, Carlo Rovelli, and many others, starting in the late 1980s and early 1990s, loop quantum gravity aspires to provide the geometric framework for a complete quantum theory of gravitation. It replaces continuous space-time with flexible networks of connections—something like a child's construction kit with sticks and clips—and uses the set of possible configurations as the basis of a quantum representation of gravity that supersedes general relativity at high energies. In doing so, it rigorously strives to fulfill John Wheeler's goal of describing quantum fluctuations of geometry on the tiniest scale: what he called quantum foam (discussed in Chapter 4).

In 1999, German physicist Martin Bojowald proposed a means of applying the theory to the universe itself, inaugurating the field of loop quantum cosmology. Later, Ashtekar and others made key contributions to its development. One of its key predictions is that quantum gravity removes the Big Bang singularity and replaces it with an amorphous start to the expanding era of the cosmos. At that point, it proposes, the universe was a quantum jumble of fluctuating geometric networks, rather than the smooth space-time associated with conventional general relativity. Because of such fuzziness, there was never a time in the past when physical quantities, such as energy density, blew up to infinity—as in the Big Bang theory. Moreover, that foamy phase was preceded by an era of cosmic contraction. One might imagine squeezing a large foam ball into a tiny glob, letting go, and watching it spring back to its original size. Thus, there was a bounce—with cosmic contraction turning into universal expansion—rather than a bang.

As with Conformal Cyclic Cosmology, loop quantum cosmology makes predictions about the CMBR. In 2015, physicist Ivan Agullo of Louisiana State University calculated that the transition from shrinking to growing would leave telltale stretch marks in the CMBR frequency spectrum. He suggested that such imprints would make their presence known through anomalous skewing of spectral data, such as that collected by the Planck satellite. Such an aberration would be subtle, and seen only through statistical analysis. However, in 2023, researchers Bartjan van Tent, Paola C. M. Delgado, and Ruth Durrer published the results of a meticulous study of the Planck data that appeared to rule out such an effect. With great confidence in their statistical analysis, they framed their findings as practically ironclad.

"It is of course possible that this might be evaded by some very exceptional, faster than exponential decay of the [effect of a bounce on the CMBR spectrum]," they write. "Nevertheless, ours appears to be a quite solid conclusion."[12]

In 2018, physicists Latham Boyle, Kieran Finn, and Neil Turok proposed yet another Big Bounce cosmology, involving a Big Bang split between our expanding universe and another "antiuniverse" traveling backward in time. They designed their model, in part, to help resolve the glaring discrepancy between the ease of producing antimatter (oppositely charged counterparts of ordinary matter) in particle physics processes—suggesting a fundamental matter-antimatter symmetry—and the overwhelming abundance of matter over antimatter in the observable universe. While particles and antiparticles turn up equally in many laboratory processes, such as electrons and positrons (their positively charged counterparts) emerging in pairs, astronomers never observe galaxies, stars, or other celestial bodies made of antimatter. (Note that dark matter is something completely different.)

One way theorists represent antiparticles, following Richard Feynman's methods, is to imagine them as particles moving backward in time. Therefore, the creation of a particle-antiparticle pair can be framed as two particles moving in opposite directions through time: one into the future; the other into the past. The backward time travel is not taken seriously, however. Rather it is simply a tool for calculation.

In Boyle, Finn, and Turok's model, the Big Bang itself represents such a splitting event. Two cosmic entities emerge from the void. One is our own universe, moving forward in time, mainly composed of matter. The other is an antiuniverse, moving backward in time, primarily antimatter. Thus, the global cosmos—a kind of multiverse of paired space-time entities—preserves a satisfying symmetry between matter and antimatter. While in their paper the team made no specific predictions for the CMBR, they predicted a new kind of massive neutrino that would coil in the opposite direction to ordinary neutrinos: right-handed instead of left-handed. So far, such right-handed, massive neutrinos have yet to be found.

With scant evidence to support them, Big Bounce cosmologies, including the Cyclic Universe, Conformal Cyclic Cosmology, loop quantum cosmology, models with an antiuniverse, and other schemes, remain minority positions. The bulk of the cosmology community continues to line up behind the notion of a hot Big Bang, followed by a brief interval of inflation, which seems best able to explain the CMBR profile found by both the WMAP and Planck satellites.

Nevertheless, the prospect that inflation has led to a vast, and possibly infinite, multiverse of which our universe is but a tiny and perhaps infinitesimal segment is jolting. Our longing to understand the working of the cosmos has led us to the prospect that we might be able to map out only the tiniest fraction of everything. That is humbling, to say the least. Likely it will continue to inspire other theorists to develop alternatives to eternal inflation and the bubble multiverse that offer the possibility of being able to chart reality more fully.

Eternal inflation aside, string theory offers its own labyrinthine multiverse with which to contend—the set of all Calabi-Yau space configurations. Curiously, the brane world cyclic cosmologies proposed by Steinhardt, Turok, and their collaborators are connected with string theory and M-theory predictions. Thus, even if those cosmologies ultimately prove correct, a multiverse could pop up in other ways. Avoiding any kind of multiverse these days seems a tricky proposition indeed.

Cyclic models continue to be a lure for those uncomfortable with complete cosmic extinction. Imagining the end of everything is absolutely frightening. Yet without the possibility of renewal, the ceaseless growth of entropy mandates a bleak ultimate future. Without usable energy, the final stages of the universe would be lifeless.

On a personal level, the passage of time can similarly be scary—particularly in facing the losses due to aging and the fear of making irreversible mistakes. We mourn the relatives and friends who are no longer with us and pine for past relationships. The idea of traveling back in time to rectify errors, spend time with those now deceased, witness and possibly correct historic events, and so forth thereby entices.

Long discussed in tales of speculative fiction, time travel remains an intriguing prospect. In recent decades, thanks to the work of gravitational theorist Kip Thorne and others, the notion has increasingly appeared in the pages of scientific journals. The concept connects with multiverse ideas in a number of ways. First, general relativity does not readily distinguish between movement in space and the flow of time. Unlike special relativity, which excludes past-directed journeys and interactions for ordinary particles (aside from imagining such for antiparticles, as Feynman suggested, to simplify calculations) due to preserving the order of causality, general relativity does not seem to forbid backward time travel. Stephen Hawking, in his Chronology Protection Conjecture, suggests ways general relativity might place obstacles to past-directed journeys, but others have found possible loopholes.

One potential conduit for backward time travel, as shown by Thorne and others, involves hypothetical journeys through traversable wormholes. Such, if somehow constructed or found, could potentially be used for temporal, as well as spatial, voyages. In either case, they might be employed to connect with realms beyond observability—one definition of a multiverse.

Moreover, backward time travel, if enacted in an effort to change past events, might well result in paradoxical situations, such as warning one's younger self never to travel in time. When the younger version heeds that advice and avoids such a temporal excursion, it would be unclear who actually issued the warning. One way of avoiding such paradoxes would be mandating that all backward-in-time loops must be self-consistent. Another hypothesis—often employed in science fiction—is to postulate that reality branches if past events are altered. The set of all branching timelines would be yet another type of multiverse. The element of bifurcation would lend it some resemblance to the Many Worlds Interpretation, yet the cause would be very different. Instead of quantum superposition, the time travel multiverse would be due to human intervention into past occurrences.

Speaking of the MWI, yet another way time travel pertains to multiverse ideas is David Deutsch's proposal that time doesn't actually flow. Rather, in a scheme that draws from Everett's hypothesis and other sources, all events in space and time constitute island universes. Our minds connect them in ways that obey the laws of physics. Therefore, our personal sense of the flow of time would be a "persistent illusion" as Einstein so aptly expressed.[13]

In short, multiverses and time travel schemes have considerable overlap. They each also face skeptics' critiques that they are not real science in the same way as, for instance, generating magnetism from an electric coil and measuring its field strength. However, with many fundamental theories at an impasse, multiverse advocates offer credible cases, and so do those who ponder cycles of time. Summarily dismissing speculative models that endeavor to solve formidable problems could turn out to be apt, or conversely represent a grave mistake in the way Einstein's reluctance to accept quantum mechanics turned out to be. Cautious openness to various options until testing or theoretical considerations rule them out completely, therefore, seems the prudent path forward.

By making a round trip on a rocket ship in a sufficiently wide curve, it is possible in these worlds to travel into any region of the past, present, and future, and back again, exactly as it is possible in other worlds to travel to distant parts of space. This state of affairs seems to imply an absurdity. For it enables one . . . to travel into the near past of those places where he has himself lived. There he would find a person who would be himself at some earlier period of his life. Now he could do something to this person which, by his memory, he knows has not happened to him.

—Kurt Gödel, "A Remark About the Relationship Between Relativity Theory and Idealistic Philosophy," presented to Albert Einstein on his seventieth birthday

You are cordially invited to a reception for time travellers hosted by Professor Stephen Hawking, to be held in the past, at the University of Cambridge, Gonville and Caius College, Trinity Street, Cambridge . . . 12:00 UT 28 June 2009, no RSVP required.

—Time travelers party invitation sent by Stephen Hawking

CHAPTER EIGHT

THE TIME TRAVELERS PARTY

Kip Thorne, Stephen Hawking, and the Prospects for Temporal Voyages

It is a sad fact of life that almost everyone, at some point, witnesses tragedy. Accidents, illnesses, and violent encounters often leave lasting trauma. Grief may be compounded by guilt if someone wonders if they could have done anything differently to avoid the heartbreaking event.

Although physics aspires to be an objective science, it is inevitable that emotionality sometimes drives preferences for models of nature—especially if experimental evidence to distinguish certain choices is lacking. One of the drivers for various multiverse and cyclic universe ideas is humanity's deep-seated wonder about alternative histories and the question of whether or not things could be different—for better or worse—elsewhere in a multi-stranded cosmos or, alternatively, in other cycles of time.

General relativity, with its bevy of offbeat solutions, has been a key driver of speculations about alternatives. Ironically, as noted, Einstein hoped studies of its properties—and of unified field theory extensions that incorporated all of the forces of nature—would converge on a single, inevitable model of reality. Instead, along with quantum theory, it has been a generator of numerous surprises about what strange theoretical possibilities lie out there. Rather than agreement on the properties of a simple, solitary universe, it has enabled ever-growing speculation about a vastly complex multiverse.

In March 1949, when Einstein celebrated his seventieth birthday, likely the last thing he wanted was drama. In his later years, he often expressed his need for peace and quiet. One of his dearest friends and colleagues, Kurt Gödel, decided to honor him with a gift, which at face value would maintain serenity: a scholarly paper on relativity and philosophy. Einstein was pleased, no doubt, by the gesture. Yet the contents of the paper were hardly calming. Gödel reported the discovery of an exact solution of general relativity, representing a rotating universe, that would theoretically allow astronauts to travel into the past. Such time travel, if possible, could lead to the paradoxical situation that a space traveler would be able to meet himself in the past and change the course of his life. Yet the future self would not recall such an encounter in his own past. What an enigma for Einstein to wrap his seventy-year-old brain around!

Luckily Einstein was not too concerned that his quest for a unique universal timeline was for naught. While disturbed by the implications of backward time travel, he wasn't going to worry about a hypothetical construct unless it was shown to be viable. Undaunted, he would continue to pursue unification until his death.

Einstein was right not to fret. Gödel's scenario depended on many assumptions that turned out either to be invalid or extremely difficult to meet. It presumed that galaxies had an unbalanced overall rotation, like ballet dancers mostly swirling in the same direction. However, no evidence has been found for such a net universal spin.

Gödel's solution also ignored the expansion of the universe. Few doubted by 1949 that the recession of galaxies, as indicated by their redshifts, meant that space was growing. Yet his model spun but didn't enlarge.

Finally, Gödel assumed space voyagers would be able to circumnavigate the universe, go backward in time, and then return safely to

Earth quickly enough to arrive some years before their departure. He freely conceded, though, that the idea of such extraordinary fast and lengthy travel was far-fetched. Not only would that need to be timed just right, but the astronauts would need to travel billions of light-years and remain healthy and alive. That was incredibly presumptuous—far more science fiction than science—especially because, at the time, human spaceflight had not even been attempted. He explained, however, that he was simply making a philosophical case for the paradoxes of time travel.

Nevertheless, that work, and Gödel's subsequent published articles where he presented the mathematical derivation and implications of his rotating solution, helped inspire a genre of theoretical research on the question of scientific time travel. Rather than the whole universe spinning around, though, the newer ideas focused on objects within the universe that might contort space-time enough to allow journeys into the past. Inevitably, time travel paradoxes arose—similar to those raised by Gödel—and some speculative thinkers brought up multiverse ideas for avoiding them, such as the notion that any change to the past might lead to branching universes. For every cosmological or quantum conundrum, these days, there seems to be a kind of multiverse offering a solution.

APPROACHING THE SPEED OF LIGHT

While time travel into the past raises philosophical and practical dilemmas, time travel into the future turns out to be relatively easy. Recall that Einstein's special theory of relativity informs us that the closer anything travels to the speed of light the slower its clock ticks relative to an observer who is not journeying with it—for example, remaining on Earth. Such time dilation has been tested by means of ultra-precise atomic clocks on high-speed aircraft. Board any fast-moving plane, as it turns out, and you've leapt a tiny bit into the future relative to those on the ground, and thus added a minuscule amount of time to your life (as measured by moments since birth). Arguably such relativity-enhanced longevity, however, would be more than negated by the effects of airline food and stale, circulated air. Note that time dilation slows down the decay of elementary particles far more notably. They are lucky that way; relativity increases the lives of lightweight things such as particles much more dramatically. And they don't have to suffer from warmed-up airline pasta.

Backward time travel is forbidden in special relativity, though, except for hypothetical faster-than-light particles called tachyons. Weirdly enough, Einstein's formula for time dilation flips from a positive to negative time interval if an object's speed exceeds that of light. However, it is impossible for a sub-light-speed particle to accelerate enough to reach light speed, let alone surpass it. Only a particle that is already faster than light might remain that way. They'd be moving backward in time forever. Despite the OPERA (Oscillation Project with Emulsion-tRacking Apparatus) group's 2011 claim of observing faster-than-light neutrinos (extremely lightweight, electrically neutral particles), later retracted due to a timing error, no tachyons have yet been found. Therefore, it could well be the case that all particles with mass move slower than light, and travel only into the future, not the past.

In general relativity, on the other hand, the space and time axes at one point might be twisted relative to that of another point, like a grove of willow trees bending in the wind. In certain cases—such as in Gödel's rotating universe—the time axes can be lined up in a closed loop, like the needles of a hedgehog, so that the future of one matches the past of the next, theoretically allowing backward time travel. Such situations are known as Closed Timelike Curves (CTCs).

In 1974, Tulane physicist Frank Tipler proposed a mathematically simple example of a CTC in his paper "Rotating Cylinders and the Possibility of Global Causality Violation." Known as a Tipler cylinder, it involves an infinite, rotating rod in space, massive enough to distort space-time with its twisting motion, in a process called "frame-dragging." Tipler calculated that it would create a CTC, allowing past-directed time travel when circumnavigating it. Astronauts attempting such a journey wouldn't have to circle around the whole universe, just the rod. The first scientific time travel idea advanced since Gödel (aside from speculations about tachyons), Tipler cylinders soon became a staple of speculative literature. For example, in 1977, science fiction writer Larry Niven published a short story about time travel with the same name as Tipler's article.

SAFE AND RAPID WORMHOLE TRANSIT

While much science fiction is based on theoretical physics, occasionally literature returns the favor and inspires scientific ideas. A perfect symbiosis between the two came about in the early 1980s, when astronomer-turned-novelist Carl Sagan was researching his fictional work *Contact* and turned to his friend physicist Kip Thorne for advice. Sagan wished to devise a way the main character of his novel could journey quickly to a remote civilization in space. He knew that there had been considerable discussion of how Schwarzschild black holes might be able to connect with other objects: additional black holes, or perhaps even their hypothetical opposites that spew energy rather than absorbing it, dubbed "white holes." Those connected bodies could potentially be very distant from each other in ordinary space—perhaps even in another part of our galaxy, or other galaxies altogether. Such a mechanism could be a conceivable shortcut to a faraway, inhabited world, Sagan pondered, at least in speculative fiction. However, he read that astronauts attempting to travel through such links could possibly end up in danger. He asked Thorne to clarify.

Thorne's extensive knowledge of the wildest solutions in general relativity, from black holes to gravitational waves, made him exactly the right person to ask. He had been fascinated by astronomy since a young age, while growing up in Logan, Utah. Logan lies in a valley that gets a lot of snow in the winter, so as a boy Thorne wanted to be a snowplow driver. At the age of eight, however, his mother brought him to a talk about the solar system. Five years later, reading a popular book by physicist George Gamow, who helped develop the Big Bang theory, sealed the deal; Thorne wanted to be an astronomer.

After undergraduate work at Caltech, Thorne began a PhD program in physics at Princeton. There, his mentor, John Wheeler, introduced him to wondrous objects in general relativity, such as black holes, geons, wormholes, and so forth. Later Misner, Thorne, and Wheeler jointly authored an innovative textbook, *Gravitation*, that remains a classic. Thorne's subsequent work in gravitational wave detection, along with Rainer Weiss and others, culminated in the successful LIGO (Laser Interferometer Gravitational-Wave Observatory) detectors, and Nobel Prizes for both of them in 2017, along with former LIGO director Barry Barish, for discovering the first gravitational wave signals.

In responding to Sagan, Thorne pointed out that venturing into a black hole, with the hope of finding a spatial shortcut to elsewhere, would not be a good idea. Astronauts would be stretched like taffy due to its intense gravitational field, bombarded by lethal energy released by infalling matter, and accelerated to a dangerous level like the most unsafe thrill ride imaginable. Moreover, even if they could somehow survive those perils and find a wormhole connection to another part of the galaxy, its throat would be unstable, and shut off immediately upon entrance. There would be no way of making it through. In short, the mission would be doomed to failure, and almost certainly to death.

That dialogue motivated Thorne to think of a solution to Sagan's query. He assigned Michael Morris, one of his graduate students at Caltech—where he had become a professor—the task of designing a safe, traversable wormhole, if that were at all possible. Morris took on the task with great enthusiasm, and worked with Thorne to find a general relativistic solution with stable wormhole throats, swift, secure, and comfortable passage through them, and minimal risk when entering and exiting. Indeed, they found a traversable wormhole that would meet those specifications.

There was one catch, though: the wormhole would have to be built with enormous quantities of matter—perhaps comparable to the heft of galaxies—and, at least partly, with an unknown antigravity material with negative mass, dubbed "exotic matter." No known astronomical or terrestrial objects have masses less than zero. There are ways, however, in quantum field theory to construct states with negative overall energy, and hence negative mass. Conceivably, a highly advanced civilization could thereby mine the quantum vacuum in negative energy regions to collect exotic matter for wormhole construction. Then it would be faced with the additional challenges of finding the gargantuan amount of ordinary masses needed and building the wormhole according to safe and speedy specifications.

In 1987, Morris and Thorne published a paper with their findings in a pedagogical physics journal, *American Journal of Physics*. Their work inspired other physicists to try their hand in developing wormhole models. In particular, New Zealand physicist Matt Visser soon found alternative solutions, which he determined to require a smaller percentage of exotic matter. Though constructing traversable wormholes remains far beyond our capabilities, and may ultimately prove to be impossible, Visser's results offered a welcome step forward.

In the 2014 movie *Interstellar*, Thorne, as co-producer and scientific consultant, explored the idea of traversable wormholes in fiction in far greater detail than Sagan had attempted. The film uses them as a plot device for astronauts to explore other worlds in our universe to assess their habitability. Yet if traversable wormholes are feasible, they could well link to other universes—that is, otherwise-disconnected parts of space—instead of our own universe. Conceivably, therefore, by offering passageways between disparate cosmic enclaves, such a wormhole network would represent yet another kind of multiverse. Note though that the theory of wormholes is not yet developed enough to determine which spatial regions they'd connect. One might only speculate.

TIME TUNNELS

One unforeseen consequence of traversable wormhole theory revived debates about whether or not Closed Timelike Curves exist in nature. Along with Ulvi Yurtsever, Morris and Thorne discovered that, by accelerating one of the mouths (entrances and exits) of a navigable wormhole relative to the other mouth, one could engineer a situation that creates a CTC and hence seems to permit backward time travel. Specifically, the accelerating mouth would zoom outward and then boomerang back relative to the first, engendering a situation akin to the so-called "twin paradox" thought experiment in special relativity.

In that hypothetical scenario—which turns out to be self-consistent, rather than a true paradox—we envision twins with different career choices. Picture Teresa, a London corporate executive who has never stepped out of the UK, let alone Earth, and her twin sister Celeste, who left London as a teenager, joined the Space Academy, and became an accomplished astronaut. Each sister is age twenty-five at the start of Celeste's maiden interstellar voyage—and of Teresa's wedding to another executive named Tim. The ceremony happens to be exactly the same day as the launch, leading to much animosity between the siblings. It was an unfortunate scheduling error for sure.

Now suppose Celeste travels on what, according to her spaceship's clock, is a five-year voyage (2.5 years outward and 2.5 years back to Earth). She is traveling consistently at 99 percent of the speed of light, for both the outward and return journeys. According to time dilation in special relativity, the duration of her voyage according to a terrestrial observer, such as Teresa, would be more than thirty-five years instead. Celeste, now thirty according to her own timeframe, returns to a scolding from sixty-year-old Teresa, who has a long list of events in her life that she thinks would have been better if her sister had shared them.

If we replace Celeste's spaceship with one of the mouths of a wormhole, and posit a stationary second mouth that remains close to Earth (with Earth somehow protected from the enormous gravitational effects of the wormhole's huge mass), one might envision how the backward time travel envisioned by Thorne and his team would work. The moving mouth would experience time dilation and age much slower than the stationary mouth. Therefore, by the time it returned to Earth's vicinity, decades or even centuries might have passed. Yet it would still be connected with the stationary mouth positioned near Earth so long before. Consequently, by entering the moving mouth in the future—the time of its return—traveling through the wormhole's throat, and popping out of the stationary mouth, an astronaut might journey backward in time to the era in which the wormhole was established.

Hence, if, unknown to Celeste, an advanced civilization set up a wormhole at the time of her original launch, with one mouth remaining near Earth and the other trailing her spaceship (cloaked by invisibility shields, perhaps, to prevent immediate detection) and returning with her more than thirty-five years in Earth's future, she'd have a way of making it up to her sister. By traveling through the mouth that had been moving, she could pass through the wormhole's innards and exit the stationary mouth at the time of its creation. She'd then be able to help celebrate Teresa and Tim's wedding—not at the ceremony, but perhaps throwing a party sometime later. The wormhole's CTC, allowing backward time travel, would thereby help save Celeste and Teresa's sisterly relationship.

In 1991, Princeton physicist J. Richard Gott offered another example of a CTC. His time loop would be created by a pair of "cosmic strings": hypothetical colossal tubes of energy left over from the Big Bang. They are not to be confused with the minuscule vibrating threads of string theory. Astronauts would circle around the pair to take advantage of the CTC and venture into the past. They could not journey further back than the creation of the pair of strings, however. Gott's proposal, along with the work of Thorne's group, helped fuel growing interest in the question of whether or not the laws of physics permit backward time travel.

CLOSED LOOPS AND BRANCHING TIMELINES

Forward time travel generally does not create paradoxical situations. The "twin paradox" has been called that because of the misconception that both twins are on equal footing, and should each experience the other one aging slower due to time dilation. Therefore, according to that misunderstanding, they'd each see the other one as younger (or older) than they are; clearly an impossibility, unless they are delusional. In fact, there is no paradox because their situations are different: while the Earthbound observer always remains in the same inertial frame, the space traveler switches between two different inertial frames as she travels outward and back again, accelerating (by turning around) between the two. Special relativity thereby singles out the astronaut twin as the one experiencing time dilation relative to the terrestrial twin.

In contrast, backward time travel might readily engender genuinely self-contradictory circumstances. The Grandfather Paradox, in which someone goes back in time, murders their grandfather, and prevents their own birth, offers a particularly gruesome example of such a philosophical quagmire. If the time traveler were never born, how could he go back in time and carry out his heinous act? Therefore, the grandfather would be spared, and the time traveler would actually be born—glaringly contradicting the cancellation of his birth his evil deed should have caused.

Or what if, in our tale of Teresa and Celeste, the latter used a wormhole that was built much earlier, and thereby offered travel even further back in time? Once back in time she convinced her former self not to miss Teresa's wedding, so the younger Celeste ceded her role as commanding officer of a spaceship and remained on Earth. Not only would there be two Celestes at the same time—leading to a difficult decision for Teresa about whom to name as her maid of honor—but then it would be unclear how the time-traveling Celeste came to be, if the original Celeste never left Earth.

In 1990, as an attempt to avoid such knotty conundrums, Russian physicist Igor Novikov joined with Morris, Thorne, Yurtsever, and several others to propose what became known as the "Novikov Self-Consistency Principle." The idea is that any CTCs that are fully consistent in their description of events are allowable forms of backward time travel. Nature blocks any potential paradoxes (such as the Grandfather Paradox) from occurring by wrapping up loose ends into a totally coherent strand. For example, if someone tries to go backward in time to assassinate Hitler and prevent the Second World War, he might find, on arrival in Nazi Germany, that he is mistaken for a British spy and locked up indefinitely under solitary confinement. Newspapers would simply report that an anonymous spy was imprisoned for life, and history wouldn't change.

On the other hand, if self-consistency didn't hold and changing the past were indeed possible, some speculate that a backward time traveler might find himself stuck in another branch of reality. Ray Bradbury's classic tale "A Sound of Thunder" contemplates such a strange scenario of splitting timelines. In it a time travel safari takes a cowardly hunter from the future back to the age of the dinosaurs. When he panics, steps off the designated path, and crushes a butterfly under one of his shoes, he inadvertently sets off a chain reaction that eventually places him and his companions in a new branch of reality. In it the English language is somewhat different, and an important, close presidential election has gone the wrong way. In the altered timeline, an autocratic leader takes power instead of the original outcome of a good guy winning. Such a history-changing time travel scenario would constitute a multiverse of branching timelines.

Other tales with such bifurcating realities stand out as thought-provoking examples of the genre. In the 1939 novel *Lest Darkness Fall* by L. Sprague de Camp, an archeologist from the present day suddenly, during a thunderstorm, finds himself back in ancient Rome, during an era in which its civilization was under threat. By bringing that society modern knowledge, he manages to stave off its fall into chaos. Thus, he sets the world on a different timeline, in which technological progress occurred far more rapidly.

Ward Moore's *Bring the Jubilee*, published in 1953, involves a time traveler carrying out actions in the past that affect the outcome of the Civil War. *The Man in the High Castle*, by Philip K. Dick, envisions a world in which the Axis powers defeated the Allies in World War II, and carved up North America into various sectors of influence: German, Japanese, and neutral. More recently *The Years of Rice and Salt* by Kim Stanley Robinson ponders a scenario in which the Black Death wiped out European society, leaving only a tiny percentage of survivors. Consequently, in that parallel reality, Europe became dominated by its Asian and Middle Eastern neighboring cultures.

Larry Niven's "All the Myriad Ways," first published in *Galaxy* magazine in October 1968, and bearing some (likely coincidental) resemblance to the MWI, begins with the intriguing passage:

> There were timelines branching and branching, a mega-universe of universes, millions more every minute. Billions? Trillions? . . . The universe split every time someone made a decision. Split, so that every decision ever made could go both ways. Every choice made by every man, woman, and child on Earth was reversed in the universe next door.[1]

One of the key themes of Niven's story was that in a "mega-universe" of alternatives, in which every choice we make in our timeline might differ radically in other timelines and somehow we'd know that, we'd have little motivation to lead moral lives. A pacifist who works at a soup kitchen helping feed homeless refugees would almost certainly be distressed to learn that his counterpoint in another branch of reality is a war criminal. Perhaps, with that knowledge, he'd become cynical and give up his charitable work. Niven's imagined society thereby is plagued with violent crime, including homicides and suicides. Regarding the latter, one is reminded of Liz Everett's tragic decision to take her own life, writing in her suicide note that she hoped to join her father (Hugh Everett) in a parallel world.

DOES NATURE PROTECT ITS OWN HISTORY?

Branching realities aside, yet another possibility is that reality forbids backward time travel altogether, as suggested by Stephen Hawking. While a good friend of Thorne, Hawking seemed to enjoy the times when they engaged in cordial disagreements about matters of physics—and time travel proved to be no exception.

Sometimes, the two placed bets on whether something was true or false. For instance, in 1974 Hawking wagered that a certain celestial object, Cygnus X-1, was not a black hole, and Thorne bet that it was. Thorne proved right. In fact, there isn't a recorded incident of Hawking winning a bet against Thorne, but he seemed to have fun wagering anyway.

For time travel, Hawking differed with Thorne's team but apparently decided not to bet—thinking perhaps that it would be virtually impossible to prove his case resolutely. Nevertheless, Hawking surmised that because we have never encountered any time travelers from the future, such past-directed voyages are likely impossible. Otherwise, what would stop them from visiting, if only just to be tourists of the past?

In 1992, Hawking published a paper "Chronology Protection Conjecture," offering a case that the underlying principles in general relativity would block the possibility of backward time travel. Any attempts to do so—via wormholes, for example—would lead to those passages closing in on themselves, like a harrowing entrapment scene in an Indiana Jones movie.

Hawking concluded with a reference to the lack of visitors from the future. As he wrote:

> There seem to be theoretical reasons to believe the chronology protection conjecture: *The laws of physics prevent the appearance of closed time-like curves.*
>
> There is also strong experimental evidence in favor of the conjecture from the fact that we have not been invaded by hordes of tourists from the future.[2]

To emphasize his point, Hawking later advertised a party at Cambridge for time travelers from the future. Theoretically, if backward time travel were possible, anyone living in centuries to come could venture back in time and attend. Alas, nobody showed.

AN ARCHIPELAGO OF MOMENTS

In 1997, David Deutsch published *The Fabric of Reality*, an ambitious effort to explain the nature of time by means of modifying the Many Worlds Interpretation. Along with Hugh Everett, Deutsch listed philosopher Karl Popper, mathematician Alan Turing, and naturalist Charles Darwin as his key influences in writing the book. In it, Deutsch sketches a multiverse of disparate places and moments. Anything allowable by the laws of physics would present itself as islands in that eternal reality, linked together by such principles. For example, if energy conservation allows for a set of ten possible outcomes of a certain quantum transition at a certain time, there would be ten different outposts in space-time representing each of those alternatives. If certain outcomes were more likely than others, the percentage of isolated results representing those possibilities would be proportionally greater.

In Deutsch's multiverse, space and time are on completely equal footing. Therefore, just as space doesn't flow, time doesn't either. Rather, our minds link from outpost to outpost in that multiverse according to our own free will, as long as our connections in space-time obey the fundamental principles of physics. Those laws would prevent us, for example, from walking through solid walls with any degree of likelihood.

Without the flow of time, our sense of the past versus the present stems from our perceptions—with the former seeming more remote in terms of retrieving information about it and the latter appearing more immediate. An analogy would be a computer with a working memory and a long-term storage drive with various sectors. Ease of access to certain memories would be associated with the present or immediate past. Cumbersome retrieval would correspond to the distant past.

As Deutsch explained:

> We do not experience time flowing, or passing. What we experience are differences between our present perceptions and our present memories of past perceptions. We interpret those differences, correctly, as evidence that the universe changes with time. We also interpret them, incorrectly, as evidence that our consciousness, or the present, or something, moves through time.[3]

In other words, according to Deutsch, time's flow, though strongly felt, would be an illusion. No matter how much we sense the passage of minutes—while waiting for a bus, for example—reality would be a persistent web of connections. The brain often creates false sensations, such as the sense that a tile floor is freezing cold, when rather it is simply a good conductor of heat. Therefore, it is not out of the question that our keenly felt sense of time might purely be a feeling rather than a fact. On the other hand, there seems no obvious way to prove such a provocative hypothesis.

Given the plethora of alternative realms, Deutsch's multiverse would help resolve potential time travel paradoxes. Travel back in time (presuming the laws of physics allow that), attempt to change the past, and you'd simply find that your perceptions and memories are linked to a different web of connections. There would be no contradiction.

Illusion or not, time's passage is what we live. Often, unfortunately, it comes with a sense of loss. We wonder, where did our youth go? Is it out there somewhere? If so, is there any way we might access it again, such as through cycles of time (as Nietzsche pondered), duplicate worlds in space (as Blanqui contemplated), or personal time travel into the past?

Such existential questions drive interest in alternative history ideas, along the lines of the science fiction stories mentioned. They also inspire speculation about doppelgängers and evil twins, each of whom might lurk in hidden enclaves of reality. If we spend our lives being painstakingly polite and exceedingly cautious, we might wonder if there might be another version of us in a parallel universe who is brazen and bold—and all the better for it. The grass is always greener, as the saying goes—so we ardently try to gaze over the fence that separates our timeline from other possibilities. Consequently, many of our personal multiverse fantasies stem from our curiosity about other roads we might have taken.

What do such subjective ruminations have to do with the MWI, disconnected and/or wholly unobservable regions of space, bubble universes, the string landscape, and other scientific multiverse speculations? Very little, as it turns out. Except for the MWI, which postulates different experiences of random quantum outcomes (normally not particularly meaningful), none of those involve parallel human lives. Rather they concern the lofty domains of high-energy physics and cosmology.

In contrast, what does most of the public think of when they imagine a multiverse? Alternative histories and paths not taken in private lives come to mind. They picture very different meanings of the term for sure.

A multiverse model's appeal to scientists is finding a broad-enough arena, narrowing it down, and using the honed version to support a comprehensive, self-consistent description of all observed natural phenomena. For the public, though, visions of their superhero (or otherwise luckier, healthier, more effective, and so forth) selves, living happily in parallel realities, might come to mind. That disconnect offers a blessing and a curse. Popular interest helps boost funding, but blurring the line between fantasy and science potentially casts doubt on a subject some theorists see as the optimal way forward.

[Feynman] had a series of lectures on science at Cornell in which he talks about how to avoid fooling yourself. And what you have to avoid are theories which are so flexible that, after you make an observation, you say, "Oh, yeah, I know how to fix the theory to explain that. I turn this knob, I turn that knob . . ." And the multiverse is kind of the ultimate limit where all the knobs have been turned. So everything is possible and therefore nothing you measure or any combination of things you measure can be inconsistent. But that means the theory also has no power, because the point of a theory is to powerfully explain something.
—Paul Steinhardt, AIP oral history interview by David Zierler, June 2020

To me the whole point of natural science is the confrontation of theory and practice, which has proved to be wonderfully productive. But I expect that the community eventually will arrive at a cosmological theory that is fully self-consistent and consistent with all available data, but the world economy will not be able to afford to test any of the predictions of the theory. Shall the community conclude that this theory must be an excellent approximation to reality? I hate the thought but expect that this is the inevitable result of the search for the "theory of everything." Would this be reality or only an invention of exceedingly clever people?
—P. James E. Peebles, personal communication to the author, May 2022

CONCLUSION

The Reflecting Pool and the Sea: Contemplating the Meaning and Purpose of the Multiverse

In childhood, we learn the extent of our limitations, as well as our abilities. That balance might change as we grow older, but we still understand that there are limits in life. No one on Earth is omnipotent, omniscient, or immortal.

Similarly, humanity, over the millennia, along with its growing capabilities, has undergone a profound awakening as to the degree of science's boundaries. While our knowledge has grown immensely, our estimated extent of the cosmos has not just kept up pace; rather, it has far exceeded our understanding. Unknown substances—dark matter and dark energy—seem to make up the bulk of the content of the observable universe. Beyond its frontiers, and onward to potentially infinite space, who knows what is out there?

The twin drives for knowledge beyond one's limitations—personal versus universal; a reflecting pool versus an endless sea—drive distinct types of interest in multiverse hypotheses. For the public, the question of "What if?" looms especially large. Permutations of familiar incidents and personalities fascinate us, such as pondering what Elvis Presley or John Lennon might have accomplished if either had lived decades longer. No wonder Frank Capra's 1946 film *It's a Wonderful Life*—envisioning a parallel world in which the movie's protagonist witnesses what it would have been like if he had never existed—remains a perennial holiday classic.

For scientists, on the other hand, contemplating multiverse schemes involves trying to resolve far broader questions about ultimate unity, and is not a step taken lightly. Decades of frustration trying to unify the natural forces, attempting to remove the role of conscious choice in quantum measurement, and striving to explain cosmic coincidences, such as the near-uniformity of the Cosmic Microwave Background Radiation, and the low, but nonzero, value of the cosmological constant, have led many theorists to reach out for novel solutions—including multiverse notions. While seemingly not directly testable, hope remains for indirect tests of such methods. Failing that, the prospect of bringing everything observable under the canopy of a comprehensive explanation serves for many as enticement enough.

Acceptance of multiverse models in science clearly is a matter of taste. Many noted figures—for instance, writer John Horgan—remain vehemently opposed to any type of multiverse, dubbing the whole notion "unscientific." In contrast, other thinkers, such as Martin Rees, astronomer royal of the United Kingdom, feel reasonably confident—given the need to fine-tune of the constants of our universe, possibly by contrasting it with others—that a multiverse exists.

As he remarked:

> About 15 years ago, I was on a panel at Stanford where we were asked how seriously we took the multiverse concept—on the scale "would you bet your goldfish, your dog, or your life" on it. I said I was nearly at the dog level. Linde said he'd almost bet his life. Later, on being told this, physicist Steven Weinberg said he'd "happily bet Martin Rees' dog and Andrei Linde's life."
>
> Sadly, I suspect Linde, my dog and I will all be dead before we have an answer.[1]

While the personal idea of alternative realities arguably attracts the most public attention, for the sake of science it remains vitally important to separate facts, theories, and pure myths. An explosion of public interest in the term "multiverse" in recent years almost certainly stems from references in fantasy—particularly the Marvel Cinematic Universe (MCU), imaginative cartoon series such as *Rick and Morty*, and arty, innovative films such as Academy Award winner *Everything Everywhere All at Once*—rather than any scientific developments.

CLOSE ENCOUNTERS OF THE DOPPELGÄNGER KIND: A LOOK AT MULTIVERSE MYTHS

Characterizing the relationship between a "multiverse" and the human experience long predates its contemporary scientific meaning. American philosopher and psychologist William James coined the term in 1895, in an essay on optimism and pessimism, "Is Life Worth Living?" In his sense of the word, a "moral multiverse" means a universe that is neither good, nor evil, but, rather, ambivalent to virtue and vice. As he wrote:

> Truly, all we know of good and duty proceeds from nature; but none the less so all we know of evil. Visible nature is all plasticity and indifference—a moral multiverse, as one might call it, and not a moral universe.[2]

Thus, according to such logic, if every good person had an evil twin, and the converse, nature would deem those duos as balanced and equitable as positive and negative charges. Given such moral equivalence, it doesn't seem such a stretch to picture that every hero has a nearly identical villainous soulmate in a parallel reality, such as the *Star Trek* crew's evil counterparts in the classic, original-series episode "Mirror, Mirror." Such an encounter is what many in the public think of when they imagine a multiverse, but there are no credible physical theories purporting such a good-evil mirroring.

A more recent, but still nonscientific, use of the term can be found in the science fiction novels of Michael Moorcock, starting around

1970. His sense of the expression was very different from that of James. Rather than a morally ambiguous cosmos, it meant a system of interacting parallel realities. In them, a character, such as the "Eternal Champion," might have different guises in various worlds, but still have an overarching personality that reveals itself in the situations encountered.

Similarly, in the recent MCU's multiverse-themed streaming series *What If . . . ?* episodes center on Marvel superheroes having different life stories in parallel realities. Episode themes have included "What if Doctor Strange lost his heart?" "What if Ultron [a highly intelligent, human-eradicating android] won?" and "What if the Avengers never formed?"

The outstanding MCU series *Loki*, created and cowritten by Michael Waldron (who also contributed to *Rick and Morty*), offers another example of the increasingly popular multiverse genre. In the various episodes, the mischievous main character meets numerous variations of Loki from parallel universes, including both male and female versions. Though it is sometimes hard to keep track of the alter egos, the series makes it clear that one of the Lokis is canonical and the others, to some extent, auxiliary.

One clever premise of the series is the existence of an organization called the Time Variance Authority (TVA) that protects the "Sacred Timeline" of the universe—a canonical history that leads to the current, relatively peaceful society—from disruption. Variants, such as the main Loki at the start of the series, try to alter events of the past in such a profound way as to disturb the timeline to various degrees. After being captured by agents of the TVA, brought to their headquarters for trial, and threatened with imprisonment or possibly even extermination, Loki reluctantly joins with TVA agents to travel through time to key moments and help seek out and stop other Variants from doing irreparable harm to the Sacred Timeline. Despite the complex plotline, the series manages to remain coherent. It leads to an unexpected ending in which the multiverse appears to descend into chaos.

CONCLUSION

In *Everything Everywhere All at Once*, Evelyn Wang, a woman who runs a laundromat and is engaged in an onerous tax dispute with agents from the Internal Revenue Service, suddenly finds herself drawn into a battle for the very future of the multiverse. Through a process called "verse-jumping," she encounters parallel universe versions of family members—including her husband, her daughter, and her father—who hail from a reality called the Alphaverse. Its name derives from its status as the first universe to make contact with others. Unfortunately, Jobu, the Alphaverse version of Evelyn's daughter Joy, has done so much verse-jumping that the process has become automatic and she has gone mad. In her lunacy, she creates a universe-gobbling vortex called the "everything bagel," and has assembled a group of cultlike followers that worship it and her. Realizing the need to stop Jobu, Evelyn learns how to acquire similar powers herself. In the process, she learns about unrealized avenues of her life decisions, including versions of herself who are a chef, a singer, and an expert in martial arts. Ultimately, in seeing all the possible paths, she realizes greater acceptance and empathy toward the world in general.

Do such fictional multiverse adventures constitute credible cinematic depictions of Everettian splitting? Do they represent dramatic renditions of speculative physical hypotheses? Not really. In essence, they are simply pure entertainment with little to do with scientific theory.

One popular myth about the Many Worlds Interpretation, and quantum physics in general, is that the superimposed blend of realities they describe has something to do with stark alternatives in human lives, such as healthfulness versus illness, wealth versus poverty, and so forth. That misconception is sometimes promoted in self-help books, some of which claim that quantum rules allow people to choose their own fate. Make a decision to think positively about health or wealth—many such books and websites falsely advertise—and quantum processes follow those inclinations, leading to success. If only that were true, but alas it seems wishful thinking at best.

The already-strange MWI splitting in which there are two different possible outcomes of a quantum process (and thus bifurcated timelines of an observer's experience of those possibilities) often becomes mis-categorized as something like "every time a person makes a choice they divide into two versions." On the contrary, the MWI eliminates free will from quantum evolution by removing the idea of researcher-induced collapse. Therefore, observers' experiences are governed by chance not choice. Each "you" that transpires just happens to be a close-to-identical witness to a quantum event, arbitrarily observing one of a range of its possible outcomes that might have subtle differences. Moreover, the near-replica versions of the experimenters after measurement could never actually meet to compare notes, let alone argue with each other.

The MWI would not produce, for instance, dueling Kirks, bearded-versus-clean-shaven Spocks (as in "Mirror, Mirror"), battling Lokis, Evelyns encountering the roads not taken, and so forth. Rather it might distinguish between near-identical versions of a scientist witnessing one type of blip versus another type of blip (representing charged particles aligned or counter-aligned with a magnet, for instance) in a detector. The truth—presuming the MWI might somehow be proven—would hardly be a nail-biter, as some fictional accounts envision.

SLIDING DOORS MOMENTS

Though having little to do with scientific multiverses, stories about alternative timelines entice us with the intriguing prospect of observing what would happen if people made different choices in their lives. Readers or viewers might pine for the experiences they didn't have because of poor decisions, or accidents, illnesses, and other chance misfortunes that disrupted their plans. For example, someone might have been dating two people or applying to two different jobs, and had to make a choice. Imagine having access to recordings of the events that would have transpired along the other paths leading from those life-altering junctions. Through fiction (but not in scientific ideas such as the MWI or bubble universes) one might vicariously probe such parallel possibilities.

In 1998, the film *Sliding Doors*, written and directed by Peter Howitt, was released. Its innovative format, depicting parallel realities with key differences and surprising convergences by means of scenes alternating between the two versions, stimulated considerable interest in the notion of chance pivotal moments in life. In it the lead character either just catches or barely misses a subway train. In the former timeline, she arrives home in time to discover that her boyfriend is cheating on her, and decides to leave him. She finds a new boyfriend, who is much better for her. In the latter, she doesn't immediately learn about her old boyfriend's cheating, and stays with him for a time, as he continues his affair with another woman. While viewers might think, at first, that the first timeline is more liberating and beneficial for the protagonist, as the film progresses disturbing events unfurl in both timelines, and eventually support the idea that the second alternative turns out better than the first—including eventually meeting the guy she was "meant" to meet.

As writer Ashley Fetters interpreted the film's message:

> According to the *Sliding Doors* philosophy, in other words, even when our lives take fluky, chaotic detours, ultimately good-hearted people find each other, and the bad boyfriends and home-wreckers of the world get their comeuppance. There's no freak turn of events that allows the cheating boyfriend to just keep cheating, or the well-meaning, morally upright soulmates to just keep floating around in the universe unacquainted.[3]

The thought-provoking themes of *Sliding Doors* inspired the expression "sliding doors moments," applied to critical junctures in life. For instance, Princess Diana's sudden decision in August 1997 to travel to Paris with her friend Dodi Fayed (and escape the critical gaze of photographers while vacationing together in Sardinia) proved fatal when tragically they were killed in a car crash after being chased by paparazzi. Hannah Paine, writing for *News Corp Australia*, called that choice to fly to Paris, rather than, for example, straight back to London, a "sliding doors moment."[4]

During the years immediately before and after the release of Howitt's film, the television series *Sliders* similarly explored the theme of parallel universes. In that show, the main characters use a wormhole to pop into alternative history timelines, such as parallel versions of the United States in which it lost the Revolutionary War. Ultimately, they are trying to return to their own reality.

Though entertaining, even those who designed hypothetical traversable wormholes and showed how they might be used for time travel have explained that the technical issues involved and the need for enormous quantities of matter would challenge even an advanced civilization's efforts to construct them. Once again, therefore, fictional and scientific accounts remain dramatically far apart.

Arguably the most entertaining and creative television rendition of a fictional multiverse is the animated *Rick and Morty* on Adult Swim, associated with the Cartoon Network. Starting as a parody of the film *Back to the Future*, with Rick as an alcoholic, egomaniac alternate version of "Doc" Emmett Brown and Morty as Rick's grandson and sidekick, at the time of this writing, the show is now in its sixth season. In virtually every episode the characters either venture to parallel universes or are visited by beings from such. Rick, Morty, and their families have numerous near-replicas in other realities. Each parallel universe has a catalog designation. According to early seasons, the name of the universe where Rick and Morty begin their adventures is "C-137." Later it is revealed that the main Rick's home universe is different from that of the main Morty's. The "137" might possibly be a reference to the numerological significance Arthur Eddington assigned to that number, as the near-inverse

CONCLUSION

of the fine structure constant. However, halfway through the first season that universe becomes "Cronenberged" (corrupted by weird creatures in the manner of director David Cronenberg's horrific creations), and Rick and Morty flee to a parallel Earth in which they take over the lives of versions of themselves who've just died in that world. Because of such switches, keeping track of the universes in the show's complex multiverse is a tricky business.

Note that in none of the scientific multiverse schemes might characters effortlessly pop in and out of various parallel universes, in the manner of Rick and Morty, as if they were visiting different lands in a theme park. Multiverse models in physics don't make travel easy, to say the least. No one has the capability to observe, let alone visit, the various neighboring bubble enclaves in eternal inflation, nor to switch places, at will, with their counterparts in the MWI. Fictional multiverse notions remain, nevertheless, a growing category of entertainment—with the science behind them twisted beyond recognition.

MULTI-TAINMENT AT THE MULTIPLEX

Film renditions of comic book characters are big business these days, attractive to distinguished actors who might wish to balance their rewarding, but relatively low-paying, roles in independent films with lucrative parts in blockbusters. The same actor might hop from Shakespeare to a cyborg villain and be lauded as well as financially rewarded in the same year. In 2003, for instance, Patrick Stewart portrayed King Henry II, in *The Lion in Winter*, and Prof. Charles Xavier, in *X2: X-Men United*, part of Marvel's *X-Men* series of movies.

What to do, however, in a commercial film franchise such as Marvel's, if an actor ages out of a role, or is otherwise no longer interested in portraying it? Or, alternatively, what if in one film a character dies or is marginalized, but the part is perfect for another film and the actor is keen on playing it? What if a villainous character becomes popular in one film, and seemed primed to become a hero (or an ambiguous hero, at least) in a starring role in another?

In the past, for each of those circumstances, audiences would have to suspend disbelief and accept that the same roles might be played by different actors (think of Darrin in *Bewitched*), or that characters might change their personalities in sequels (think of Ed Asner's Lou Grant as a comic newsroom persona on *The Mary Tyler Moore Show* and a serious newspaper editor on *Lou Grant*). If the producers wish the audience to understand the changes in the characters, or to explain how a character that seemingly died is now well, they might put the screenwriters up to the task.

Think, in contrast, of a cinematic multiverse, such as the novel direction the MCU has taken in recent years. There is no need to explain new actors and the changes in the looks and personalities of characters, given that everything is apparently possible in parallel universes. Audiences who loved Spider-Man and the villains who oppose him, such as Vulture, Electro, and the Green Goblin, might be entertained and kept on their toes by encountering somewhat different versions of those characters—good or evil—from alternate worlds. Potential plot twists magnify with no bounds, save keeping the action going in a way that the audience might follow.

As film critic Clarisse Loughrey slyly noted:

> I'd always thought that the appeal of the multiverse lay in its infinite possibility. Imagine the only limit to existence being the breadth of our own imagination—that anything we could conjure could be out there, birthed into existence in an alternative universe. Well, thank you, Marvel [Cinematic Universe], for showing me how wrong I was. It turns out that the point of the multiverse . . . isn't its creative potential. It's its cameos. A million universes could exist, and they'd all contain surprise appearances by people and things fans can hoot and holler over, before being purchased as toys on the way out of the cinema.[5]

CONCLUSION

Judging by box office receipts, Marvel's multiverse strategy has paid off handsomely. In 2021, *Spider-Man: No Way Home*, involving villains reaching familiar Earth from parallel realms, was the top-selling film, both in the United States and worldwide. It stayed in the top ten for the first half of 2022. That year, *Doctor Strange and the Multiverse of Madness* was in the top five for box office sales in the United States. Its storyline involved a girl with the mysterious power to travel from one parallel universe to another, assisting the titular character in saving the world. In 2023, rival franchise DC Comics jumped on the multiverse bandwagon with *The Flash*, featuring alternative realities with different versions of Batman, among other distinctions. The multitudes at the multiplexes seem to like their multiverses with variations on familiar characters, ample magic, and a mere sprinkle of scientific jargon.

None of those movie plots should be taken seriously, of course. They should be seen as pure entertainment that borrows some of the language of speculative science. Scientists are well aware that their carefully conceived ideas might be taken out of context and portrayed in outlandish ways. A notorious example is Disney's 1979 film *The Black Hole*, which depicts the interior of such an object as a Dante-esque hellish landscape. Despite the film's unrealistic treatment of its subject, the scientific study of black holes remains untainted. One might hope for the same with multiverse schemes.

BEYOND THE OBSERVABLE

Cinematic journeys aside, Earth's place in the cosmos is extraordinarily humble. Unlike fictional *Star Trek* voyagers, we are practically hermits, having ventured in crewed spaceflight only as far as our own moon. Yet our ambitions remain boundless. There is so much of the mappable universe left to explore by telescope or otherwise.

Researchers have estimated that the observable universe contains hundreds of billions, and perhaps up to roughly two trillion galaxies.

The Milky Way is thought to contain between 100 billion and 400 billion stars. Presuming we live in an average galaxy, contemplating the possible tally of stars in the observable universe is absolutely mind-blowing—possibly in the sextillions (1 followed by 21 zeroes) or more.

Based on data collected by the Planck satellite and other instruments about the age and growth rate of the expanding universe, as well as precise measurements of the speed of light, scientists have calculated the extent of the observable universe.[6] They find it to be a spherical bubble, some forty-five to forty-six billion light-years in radius, centered on Earth (the vantage point from where we take our astronomical measurements or launch instruments to do so). Not all astronomical measurements yield the same values of cosmological parameters, though, causing some anxiety (sometimes called the "Hubble crisis"), but mostly emphasizing the need for further testing to pin those figures down.

Exploring the volume of the observable universe's "bubble" is such a daunting task, it would seem that the scientific community has its work cut out for it for the indefinite future. Indeed, it does. But why stop there, if our models might take us even further—with the goal of explaining what we can see, based on carefully considered hypotheses about the greater cosmos?

Theories in physics lacking the possibility of direct instrumental verification (at least for the foreseeable future) often stir up considerable controversy. Some thinkers, following in the footsteps of late philosopher of science Karl Popper, believe that proper scientific theory must include falsifiable predictions. Realms beyond direct observation might well be wholly untestable (unless someone develops credible indirect ways of measuring what lies beyond the horizon of observability). Therefore, some critics argue, such hypotheses are not even science.

Critics, such as Paul Steinhardt and accomplished South African physicist George Ellis, argue that concepts such as string landscapes and eternal inflation fall outside the bounds of true experimental science, because they include vast domains that would never be measurable. Physicists, they emphasize, should focus exclusively on fully testable alternatives.

As Steinhardt said:

> [Eternal inflation] is not a useful scientific idea. . . . The only reason to pursue such an idea is because you're in love with it; you happen to like that imaginative idea. . . . But within the realm of science, you're not allowed to do that. . . . And as a scientist, I would challenge you to prove to me that there exists no other possibility. You cannot claim you've exhausted all the other possibilities, as some occasionally do in this field. . . . But then some say, "Anything . . . that disagrees with observations, I'm going to ascribe to the randomness of the multiverse." Then, that is not a scientifically useful way of thinking.[7]

Ellis, a Templeton Prize recipient who was yet another of Dennis Sciama's extraordinary students at Cambridge, and is now a professor emeritus of applied mathematics at the University of Cape Town, similarly calls multiverse notions "untestable" and not a predictive scientific idea. As he wrote in *Scientific American*:

> All in all, the case for the multiverse is inconclusive. The basic reason is the extreme flexibility of the proposal: it is more a concept than a well-defined theory. Most proposals involve a patchwork of different ideas rather than a coherent whole. The basic mechanism for eternal inflation does not itself cause physics to be different in each domain in a multiverse; for that, it needs to be coupled to another speculative theory. Although they can be fitted together, there is nothing inevitable about it. The key step in justifying a multiverse is extrapolation from the known to the

unknown, from the testable to the untestable. You get different answers depending on what you choose to extrapolate. Because theories involving a multiverse can explain almost anything whatsoever, any observation can be accommodated by some multiverse variant.[8]

However, as accomplished astrophysicist Virginia Trimble has noted, in the history of astrophysics there have been many situations where objects, once believed to be undetectable, eventually proved observable. She remarked:

> Some of the [other] "multis" were once also thought to be forever unobservable: Agnes Mary Clerke on how big, bright, and far away M31 [galaxy] would have to be if it were like the Milky Way. Of course, it is roughly that big, that bright, and that far away. And planets orbiting other stars, despite multiple false alarms, were thought undetectable by many pundits.[9]

Moreover, the power of conservation laws in physics, such as conservation of charge, conservation of mass-energy, and conservation of momentum, allows us to infer the properties of regimes we're unable to probe directly. As John Tyndall emphasized back in 1871, in his treatise *Fragments of Science*, they can be game-changers. For instance, one explanation of Hawking radiation from black holes posits that matter-antimatter pairs form near their event horizons. One member of the couple plunges inward, and the other, because of conservation of momentum, must head outward. Over time, such processes produce a net loss of mass-energy. Thus, momentum conservation permits us to deduce an important property of the unobservable region within event horizons. Because of symmetry rules, conservation principles, and so forth, lack of direct observability needn't be an obstacle to scientific progress.

CONCLUSION

THE PATH TO ULTIMATE UNITY

In recent centuries, some of the greatest triumphs in science have involved unifying natural interactions by means of basic sets of equations. James Clerk Maxwell's brilliant notion in the 1870s of combining known relationships between electricity and magnetism, along with his own contributions, into unified electromagnetism, described by a wave equation, would ultimately offer society stunning new advances including wireless communication. At the time that Maxwell made his proposal, however, the bulk of the electromagnetic spectrum in its invisible frequencies had yet to be revealed. Only optical, infrared, and ultraviolet light were already known. Fortunately, within decades, stalwart scientists would complete the spectrum, starting with Heinrich Hertz's clever proposal in the 1880s to measure low-frequency electromagnetic signals. His discovery of radio waves, measured to travel at the speed of light, confirmed Maxwell's hypothesis and proved absolutely revolutionary. One might wonder what would have happened if it took much longer—say a century—to verify all of the implications of Maxwell's unification. Would the scientific community have remained patient?

In the 1960s, the combined theoretical insights of Sheldon Glashow, Abdus Salam, and Steven Weinberg would lead to the melding of electromagnetism and the weak interaction in a quantum field theory, called electroweak unification. The theory predicted a number of new particles, including the W and Z exchange bosons and the Higgs boson. Fortunately, once again, all of its major predictions would be verified within the subsequent few decades, from the 1970s until the 2010s.

Given that gravitation, expressed as a particle theory, seems resistant to quantization, resulting in noncancelable, infinite terms, a large segment of the theoretical community continues to hold out hope for string theory and M-theory. But considering the lack of experimental evidence so far, patience is wearing thin. Further, it is hard to see how to accommodate the extraordinarily large range of string vacua without either novel mathematical methods or generous application of the Anthropic Principle to a string landscape multiverse.

In the age-old dilemma of whether the ends might justify the means, it is fascinating to wonder what would happen if a unified field theory eventually emerges with successful predictions in the measurable range, but also a gamut of unobservable consequences in energy domains far beyond the possibility of testing. On a scale of reasonability, its successes would need to be weighed against its gaps—such as, for example, the need for a multiverse. Then the physics community would need to determine, in light of possible shortcomings in the domain of observability, if the theory offered enough accurate predictions to serve as a new standard model. Even if the model offered many solid predictions, however, almost certainly there would still be naysayers because of the controversial multiverse aspect.

Intriguingly, some today who don't consider multiverse ideas true science are willing to accept many other things that we cannot *directly* observe, from the hidden vaults of Hilbert space to the sea of virtual particles in a vacuum, and from the cloaked interiors of black-hole event horizons to the opaque period between the Big Bang and the recombination era, when the light that cooled down into the CMBR was released. We still don't know what constitutes dark matter and dark energy, and the specific scalar field that triggers inflation—yet all those things are widely accepted. Therefore, one rebuttal to multiverse skeptics is to reap the benefits of such notions (for example, using eternal inflation to explain the CMBR profile, as well as the horizon and flatness problems) while continuing to give those developing potential schemes for experimental testing more opportunity for creative consideration. Does everything require immediate proposals for verification using direct evidence?

Indeed, while tests of scars of collisions between bubble universes have proved inconclusive so far, we might wish to give such experimentation more time—especially in the face of the notion of using polarization to look for such evidence. It is also possible that the CMBR imprint of primordial gravitational waves will eventually be found, and tell us much more about the inflationary era. Similarly, while evidence of supersymmetric companion particles has yet to be found at the Large Hadron Collider, perhaps the results of its recent upgrade will offer more success. In short, one needn't give up hope.

Advocates of multiverse ideas often point to the dearth of credible alternatives. Take, for example, the Many Worlds Interpretation of quantum mechanics. While it seems implausible to imagine continuous branching of an extraordinarily intricate universal wave function that spans reality, turning us into endless near-replicas, how else might we construct a quantum state of the cosmos itself, fulfilling the goal of a quantum theory of gravitation? No one can step out of that wave function, so perhaps we must be part of it. Moreover, many of the other choices in quantum measurement theory have major issues—either conceptually or experimentally. For example, von Neumann's notion that conscious observation induces collapse makes little sense in the context of physics or neuroscience. Until a scientific theory emerges that links the human brain with quantum wave function reduction, it is hard to see how such a mechanism might be justified. Spontaneous localization of quantum states is generally limited in its scope—focusing primarily on localization of position, as opposed to, for example, collapse into certain spin or polarization states. Moreover, many such models predict the release of minute amounts of radiation during physically induced collapse processes, which has yet to be detected despite careful searches.[10]

As its critics point out, eternal inflation, with its endless production of bubbles, sounds like it is extravagantly adding more to the picture than is required. On the other hand, some of the leading Big Bounce alternatives ask us to believe either that another universe is right in front of our noses, only a hair's breadth (along another dimension) away, or that a colossal cosmos can twist into a tiny one and reset its overall entropy for a new aeon. In short, while many of the multiverse explanations sound weird, some of the major alternatives seem equally bizarre.

With string theory, once again it is distressing to contemplate its intricacy. It would be great if its enormous array of vacuum states could be narrowed down considerably. Yet, particle-based models of quantized gravitation cannot be renormalized, with all infinite terms canceled out. Substituting finite strands for particles readily solves that problem. True, there are other methods, such as loop quantum gravity, that attempt to offer a solution. However, string theory's flexibility keeps it an attractive option—and a string landscape multiverse comes along with the territory.

Someday, perhaps, a theory of everything will make perfect sense, be mathematically concise, and have certain indirectly testable predictions. However, it may very well be the case that some of its assumptions—such as the existence of certain extremely high energy states and realms beyond direct observation—might never be tested directly. Should we accept such a model, or would we keep striving for more? Physics often has objective answers via equations. However, there are judgment calls that each scientist makes, involving matters of taste. The question of whether or not a multiverse with hidden enclaves constitutes acceptable science offers one such professional judgment. Every researcher who makes that call, based on the best information available at the time, deserves our respect, for it is not in any way an easy decision.

ACKNOWLEDGMENTS

This book would not have been possible without the kind support of family, friends, colleagues, and fellow scientists. As always, I greatly appreciate the excellent suggestions of my wife Felicia, and my children Aden and Eli. My father Stanley and my mother-in-law Arlene have similarly been highly supportive. I am also thankful to have long-standing friends who have introduced me to a variety of literature and culture, including Fred Schuepfer and Pam Quick, Michael and Mari Erlich, Abe Witonsky, Simone Zelitch and Doug Buchholz, Karl and Dori Middleman, Mitchell and Wendy Kaltz, Scott Veggeberg and Marcie Glicksman, Mark Singer, Elana Blum, Lindsey Poole and Greg Smith, Robert Jantzen, Kris Olson, Jeff Shuben, Boris Briker, Max R. Born, Victoria and Michael Carpenter, Cynthia Folio, Lisa Tenzin-Dolma, and Frank Cross. Thanks to Benjamin Hoffmann, and Liz and Evie Shanefield for their guidance. Musician Roland Orzabal has been a steady source of inspiration.

I'm appreciative of the support of the faculty, staff, and administration of Saint Joseph's University, including Cheryl McConnell, Jay Carter, Jason Powell, Jo Alyson Parker, Tricia Purcell, Jessie Taylor, Amanda Huan, Roberto Ramos, Sergio Freire, Piotr Habdas, Doug Kurtze, and Elia Eschenazi. The history of physics and science writing communities have similarly been supportive, including Marcus Chown, Nicholas Booth, Michio Kaku, Michal Meyer, Amanda Gefter, Faye Flam, Philip Ball, Michael Shermer, Brian Keating, David Cassidy, Catherine Westfall, Alberto Martinez, Michel Janssen, Peter Pesic, Robert Crease, Linda Dalrymple Henderson, Melinda Baldwin,

ACKNOWLEDGMENTS

Margriet van der Heijden, Bruce Hunt, Ed Neuenschwander, Donald Salisbury, and Paul Cadden-Zimansky.

Thanks to those I have interviewed or otherwise engaged in discussion with about the history of modern physics, including attitudes toward multiverse ideas—either recently or over the years—including Virginia Trimble, P. James E. Peebles, Bernard Carr, Justin Khoury, Hiranya Peiris, Kenneth W. Ford, Chris DeWitt, David Deutsch, Paul Steinhardt, Andrei Linde, Bernard Julia, Philip Pearle, and Raman Sundrum. I remain grateful for the excellent discourse I've had on this topic over the years with innovative scientific thinkers who have since passed away, including Stanley Deser, Freeman Dyson, John Wheeler, Charles Misner, Paul S. Wesson, John Smythies, Bryce DeWitt, and Cécile DeWitt-Morette. Thanks to Chaya Becker, of the Einstein Archives at Hebrew University, for her information about the photo of Albert Einstein used in this book, and approval to publish it, subject to the agreement of the American Institute of Physics Emilio Segrè Visual Archives (AIP ESVA). Thank you to photographer Niklas Björling for permission to use his photo. I appreciate the new policy of the AIP ESVA that "if the American Institute of Physics (AIP) is listed as the copyright holder of an image, you may use it freely without permission."

Guiding this project to fruition have been my outstanding agent, Giles Anderson, and the excellent leadership at Basic Books, including Lara Heimert and T. J. Kelleher—whom I thank for his excellent editorial suggestions. Thanks also to Michael Kaler, Kelly Lenkevich, Joseph Gunther, and Kristen Kim. I am truly grateful to all those who have had confidence in my research and writing throughout my career, in general, and supported the writing of this book, in particular. A multiverse of thanks to all.

FURTHER READING

Ananthaswamy, Anil. *Through Two Doors at Once: The Elegant Experiment That Captures the Enigma of Our Quantum Reality.* New York: Dutton, 2018.
Baggott, Jim. *Quantum Reality: The Quest for the Real Meaning of Quantum Mechanics—a Game of Theories.* New York: Oxford University Press, 2020.
Ball, Philip. *Beyond Weird: Why Everything You Thought You Knew About Quantum Physics Is Different.* Chicago: University of Chicago Press, 2018.
———. "The Many Worlds Fantasy." *Iai News*, issue 96, April 20, 2021. https://iai.tv/articles/the-many-worlds-fantasy-auid-1793.
———. "Too Many Worlds." *Aeon Essays*, February 17, 2015. https://aeon.co/essays/is-the-many-worlds-hypothesis-just-a-fantasy.
Barrow, John, and Frank Tipler. *The Anthropic Cosmological Principle.* New York: Oxford University Press, 1986.
Bartusiak, Marcia. *Through a Universe Darkly: A Cosmic Tale of Ancient Ethers, Dark Matter, and the Fate of the Universe.* New York: HarperCollins, 1993.
Becker, Adam. *What Is Real? The Unfinished Quest for the Meaning of Quantum Physics.* New York: Basic Books, 2018.
Brown, Julian. *Minds, Machines, and the Multiverse: The Quest for the Quantum Computer.* New York: Simon & Schuster, 2000.
Burton, Howard. *Conversations About Astrophysics & Cosmology (Ideas Roadshow Series).* New York: Open Agenda Publishing, 2022.
Byrne, Peter. *The Many Worlds of Hugh Everett III: Multiple Universes, Mutual Assured Destruction, and the Meltdown of a Nuclear Family.* New York: Oxford University Press, 2013.
Carpenter, Victoria, and Paul Halpern. "A Bridge Between Worlds: Parallel Universes and the Observer in 'The Celestial Plot.'" *KronoScope*, vol. 19, no. 2 (2019).
———. "Quantum Mechanics and Literature: An Analysis of El Túnel by Ernesto Sábato." *Ometeca*, vol. 17 (2012).
Carr, Bernard, editor. *Universe or Multiverse?* Cambridge, UK: Cambridge University Press, 2007.

FURTHER READING

Carroll, Sean. *Something Deeply Hidden: Quantum Worlds and the Emergence of Spacetime*. New York: Dutton, 2019.

Chown, Marcus. *The Magicians: Great Minds and the Central Miracle of Science*. London: Faber & Faber, 2020.

———. *The Universe Next Door: The Making of Tomorrow's Science*. New York: Oxford University Press, 2003.

Crease, Robert P., and Alfred S. Goldhaber. *The Quantum Moment: How Planck, Bohr, Einstein, and Heisenberg Taught Us to Love Uncertainty*. New York: W. W. Norton & Co., 2015.

Crease, Robert P., and Charles C. Mann. *The Second Creation: Makers of the Revolution in Twentieth-Century Physics*. New Brunswick, NJ: Rutgers University Press, 1996.

Deutsch, David. *The Fabric of Reality: The Science of Parallel Universes and Its Implications*. New York: Penguin Books, 1997.

DeWitt-Morette, Cécile. *The Pursuit of Quantum Gravity: Memoirs of Bryce DeWitt from 1946 to 2004*. New York: Springer, 2011.

Everett, Hugh. *The Everett Interpretation of Quantum Mechanics: Collected Works 1955–1980 with Commentary*. Edited by Jeffrey A. Barrett and Peter Byrne. Princeton, NJ: Princeton University Press, 2012.

Everett, Justin, and Paul Halpern. "Spacetime as a Multicursal Labyrinth in Literature with Application to Philip K. Dick's *The Man in the High Castle*." *KronoScope*, vol. 13, no. 1 (2013).

Fine, Arthur. *The Shaky Game: Einstein, Realism, and the Quantum Theory*. Chicago: University of Chicago Press, 1986.

Greene, Brian. *The Hidden Reality: Parallel Universes and the Deep Laws of the Cosmos*. New York: Knopf, 2011.

Gribbin, John. *In Search of the Multiverse: Parallel Worlds, Hidden Dimensions, and the Ultimate Quest for the Frontiers of Reality*. London: Penguin, 2009.

———. "The Many-Worlds Theory, Explained." *The MIT Press Reader*, 2019. https://thereader.mitpress.mit.edu/the-many-worlds-theory/.

———. *Six Impossible Things: The Mystery of the Quantum World*. Cambridge, MA: MIT Press, 2019.

Guth, Alan. *The Inflationary Universe: The Quest for a New Theory of Cosmic Origins*. Reading, MA: Perseus, 1998.

Halpern, Paul. *The Cyclical Serpent: Prospects for an Ever-Repeating Universe*. New York: Basic Books, 2003.

———. *Einstein's Dice and Schrödinger's Cat: How Two Great Minds Battled Quantum Randomness to Create a Unified Theory of Physics*. New York: Basic Books, 2015.

———. *Flashes of Creation: George Gamow, Fred Hoyle, and the Great Big Bang Debate*. New York: Basic Books, 2021.

FURTHER READING

———. *The Great Beyond: Higher Dimensions, Parallel Universes, and the Extraordinary Search for a Theory of Everything.* Hoboken, NJ: Wiley, 2004.

———. *The Quantum Labyrinth: How Richard Feynman and John Wheeler Revolutionized Time and Reality.* New York: Basic Books, 2004.

———. *Synchronicity: The Epic Quest to Understand the Quantum Nature of Cause and Effect.* New York: Basic Books, 2020.

———. "Time as an Expanding Labyrinth of Information." *KronoScope*, vol. 10, nos. 1–2 (2010), p. 64.

———. *Time Journeys: A Search for Cosmic Destiny and Meaning.* New York: McGraw-Hill, 1990.

Hawking, Stephen, and Leonard Mlodinow. *The Grand Design.* New York: Bantam, 2010.

Henderson, Linda Dalrymple. *The Fourth Dimension and Non-Euclidean Geometry in Modern Art.* Second edition. Cambridge, MA: MIT Press, 2013.

Husain, Tasneem Zehra. *Only the Longest Threads.* Philadelphia: Paul Dry Books, 2014.

Kaku, Michio. *The God Equation: The Quest for a Theory of Everything.* New York: Doubleday, 2021.

———. *Hyperspace: A Scientific Odyssey Through Parallel Universes, Time Warps, and the Tenth Dimension.* New York: Oxford University Press, 1994.

———. *Parallel Worlds: A Journey Through Creation, Higher Dimensions, and the Future of the Cosmos.* New York: Doubleday, 2004.

Kinney, Will. *An Infinity of Worlds: Cosmic Inflation and the Beginning of the Universe.* Cambridge, MA: MIT Press, 2022.

Kirshner, Robert. *The Extravagant Universe: Exploding Stars, Dark Energy, and the Accelerating Cosmos.* Princeton, NJ: Princeton University Press, 2004.

Kleinknecht, Konrad. *Einstein and Heisenberg: The Controversy over Quantum Physics.* New York: Springer, 2019.

Kragh, Helge. *Quantum Generations: A History of Physics in the Twentieth Century.* Princeton, NJ: Princeton University Press, 1999.

Kumar, Manjit. *Quantum: Einstein, Bohr, and the Great Debate About the Nature of Reality.* New York: W. W. Norton & Co., 2011.

Livio, Mario. *Brilliant Blunders: From Darwin to Einstein—Colossal Mistakes by Great Scientists That Changed Our Understanding of Life and the Universe.* New York: Simon & Schuster, 2013.

Martinez, Alberto. *Burned Alive: Giordano Bruno, Galileo and the Inquisition.* London: Reaktion Books, 2018.

Päs, Heinrich. *The One: How an Ancient Idea Holds the Future of Physics.* New York: Basic Books, 2023.

Peebles, P. J. E. *Cosmology's Century: An Inside History of Our Modern Understanding of the Universe.* Princeton, NJ: Princeton University Press, 2020.

FURTHER READING

Peebles, P. James E., Lyman A. Page, and R. Bruce Partridge, editors. *Finding the Big Bang*. Cambridge, UK: Cambridge University Press, 2009.

Penrose, Roger. *Cycles of Time: An Extraordinary New View of the Universe*. London: Bodley Head, 2010.

Pickover, Clifford. *Surfing Through Hyperspace: Understanding Higher Universes in Six Easy Lessons*. New York: Oxford University Press, 1999.

Rovelli, Carlo. *Helgoland: Making Sense of the Quantum Revolution*. Translated by Erica Segre and Simon Carnell. New York: Riverhead Books, 2021.

Rubenstein, Mary-Jane. *Worlds Without End: The Many Lives of the Multiverse*. New York: Columbia University Press, 2014.

Siegfried, Tom. *The Number of the Heavens: A History of the Multiverse and the Quest to Understand the Cosmos*. Cambridge, MA: Harvard University Press, 2019.

Steinhardt, Paul, and Neil Turok. *Endless Universe: Beyond the Big Bang*. New York: Doubleday, 2007.

Susskind, Leonard. *The Cosmic Landscape: String Theory and the Illusion of Intelligent Design*. New York: Back Bay Books, 2006.

Tegmark, Max. *Our Mathematical Universe: My Quest for the Ultimate Nature of Reality*. New York: Knopf, 2014.

Vilenkin, Alex. *Many Worlds in One: The Search for Other Universes*. New York: Hill and Wang, 2006.

Wallace, David. *The Emergent Multiverse: Quantum Theory According to the Everett Interpretation*. New York: Oxford University Press, 2012.

Wheeler, John Archibald with Kenneth W. Ford. *Geons, Black Holes, and Quantum Foam: A Life in Physics*. New York: W. W. Norton & Co., 2000.

Yourgrau, Palle. *A World Without Time: The Forgotten Legacy of Gödel and Einstein*. New York: Basic Books, 2004.

NOTES

Introduction: When One Universe Is Not Enough

1. Stanley Deser, personal communication to the author, May 13, 2022. "Tsuris" stems from a Yiddish word for trouble or woe.

2. In his "Participatory Universe" scheme, proposed in the 1980s, Wheeler speculated that present-day astronomers, through sheer observation of the universe's past, might be able to collapse its quantum state in a manner suitable for the eventual development of intelligent life, hence creating a closed loop in time of observer and observed. However, few other scientists have embraced that view, which seems to violate the natural order of cause and effect.

3. Niels Bohr to Wolfgang Pauli, reported in Freeman J. Dyson, "Innovation in Physics," *Scientific American*, vol. 199, no. 3 (September 1958), p. 80.

4. Bryce S. DeWitt, "The Many-Universes Interpretation of Quantum Mechanics," in Bernard d'Espagnat, editor, *Foundations of Quantum Mechanics* (New York: Academic Press, 1971).

5. Google Trends search for the term "multiverse," https://trends.google.com/trends/explore?date=all&geo=US&q=multiverse.

6. Max Tegmark, "The Universes of Max Tegmark," https://space.mit.edu/home/tegmark/crazy.html.

7. John Horgan, "The Seduction of the Multiverse," *IAI News*, issue 96, May 11, 2021, https://iai.tv/articles/the-seduction-of-the-multiverse-auid-1806.

8. Anna Ijjas, Paul J. Steinhardt, and Abraham Loeb, "Cosmic Inflation Theory Faces Challenges," *Scientific American*, February 1, 2017, https://www.scientificamerican.com/article/cosmic-inflation-theory-faces-challenges/.

9. Paul Steinhardt, interviewed by David Zierler, American Institute of Physics Niels Bohr Library and Archives Oral Histories, June 2020, https://www.aip.org/history-programs/niels-bohr-library/oral-histories/46757.

10. Virginia Trimble, personal communication to the author, May 13, 2022.

11. Alberto Martinez, *Burned Alive: Giordano Bruno, Galileo and the Inquisition* (London: Reaktion Books, 2018).

Chapter 1: Eternity Through the Stars

1. John Tyndall, *Fragments of Science: A Series of Detached Essays, Addresses, and Reviews*, sixth edition (London: Longmans, Green, and Co., 1879), https://www.gutenberg.org/files/24527/24527-h/24527-h.htm.

2. Don N. Page, "Information Loss in Black Holes and/or Conscious Beings," https://arxiv.org/abs/hep-th/9411193.

3. William Stukeley, *Memoirs of Sir Isaac Newton's Life*, unpublished manuscript, The Royal Society, 1752.

4. John Tyndall, in *Report of the Meeting of the British Association for the Advancement of Science*, British Association for the Advancement of Science (J. Murray: 1875), p. xcv, January 1, 1875.

Chapter 2: Theories from Another Dimension

1. Hermann Minkowski, "Space and Time," in Hendrik A. Lorentz, Albert Einstein, Hermann Minkowski, and Hermann Weyl, *The Principle of Relativity: A Collection of Original Memoirs on the Special and General Theory of Relativity* (New York: Dover, 1952), p. 75.

2. Abraham Pais, *Subtle Is the Lord: The Science and the Life of Albert Einstein* (New York: Oxford University Press, 1982), p. 152; Albert Einstein and Jakob Laub, "On the Fundamental Electromagnetic Equations for Moving Bodies," *Annalen der Physik*, vol. 26 (1908), p. 532, reprinted and translated in John Stachel et al., editors, *The Collected Papers of Albert Einstein, Vol. 2: The Swiss Years: Writings, 1900–1909* (Princeton, NJ: Princeton University Press, 1987), p. 508, https://einsteinpapers.press.princeton.edu/vol2-trans/343#.

3. Albert Einstein, quoted in "Einstein on Arrival Braves Limelight for Only 15 Minutes," *New York Times*, December 20, 1930, p. 16.

4. Daniela Wünsch, *Der Erfinder der 5. Dimension: Theodor Kaluza—Leben und Werk*, (Göttingen, Germany: Termessos Verlag, 2007), p. 11.

5. Varadaraja V. Raman, "Theodor Kaluza," in Charles Gillespie, editor, *Dictionary of Scientific Biography* (New York: Scribner, 1970), p. 212.

6. Theodor Kaluza Jr., *Erinnerungen*, unpublished manuscript, p. 24.

7. Theodor Kaluza Jr., interviewed on "NOVA: What Einstein Never Knew," originally broadcast October 22, 1985.

8. Albert Einstein to Theodor Kaluza, April 21, 1919, translated by C. Hoensalaers, reprinted in *Unified Field Theories of More than 4 Dimensions: Including Exact Solutions* (Singapore: World Scientific, 1983), p. 449.

9. Albert Einstein to Theodor Kaluza, April 28, 1919, translated by C. Hoensalaers, reprinted in *Unified Field Theories of More than 4 Dimensions: Including Exact Solutions* (Singapore: World Scientific, 1983), p. 451.

10. George Uhlenbeck, interviewed by Thomas S. Kuhn, Archives for the History of Quantum Physics, transcript, April 5, 1962, p. 16.

11. Albert Einstein to Max Born, December 4, 1926, reprinted and translated in Diana K. Buchwald et al., editors, *The Collected Papers of Albert Einstein, Volume 15 (Translation Supplement): The Berlin Years: Writings & Correspondence, June 1925–May 1927* (Princeton, NJ: Princeton University Press, 2018), p. 403.

12. John A. Wheeler, "Mercer Street and Other Memories," in Peter C. Aichelburg and Roman U. Sexl, editors, *Albert Einstein, His Influence on Physics, Philosophy, and Politics* (Braunscheig: Vieweg, 1979), p. 202.

NOTES

Chapter 3: Showdown in Hilbert's Hotel

1. Peter Byrne, *The Many Worlds of Hugh Everett III: Multiple Universes, Mutual Assured Destruction, and the Meltdown of a Nuclear Family* (New York: Oxford University Press, 2013), p. 32.
2. Kenneth W. Ford, remark to the author, May 11, 2023.
3. Charles W. Misner, phone interview by the author, December 6, 2015.
4. Charles W. Misner, "A One-World Formulation of Quantum Mechanics," *Physica Scripta*, vol. 90, no. 8 (August 2015).
5. Charles W. Misner, phone interview by the author, December 6, 2015.
6. Hugh Everett, interviewed by Charles W. Misner, American Institute of Physics Niels Bohr Library and Archives Oral Histories, May 1977, https://www.aip.org/history-programs/niels-bohr-library/oral-histories/31230.
7. Charles W. Misner, phone interview by the author, December 6, 2015.
8. Charles W. Misner, phone interview by the author, December 6, 2015.
9. Peter Byrne, *The Many Worlds of Hugh Everett III*, p. 138.
10. John A. Wheeler, "Assessment of Everett's 'Relative State' Formulation of Quantum Theory," *Reviews of Modern Physics*, vol. 29, no. 3 (July 1957), p. 464.
11. Bryce S. DeWitt, "Quantum Mechanics and Reality: Could the Solution to the Dilemma of Indeterminism Be a Universe in Which All Possible Outcomes of an Experiment Actually Occur?" *Physics Today*, vol. 23, no. 9 (September 1970), p. 161.
12. Hugh Everett III to Bryce S. DeWitt, May 31, 1957, reproduced on the PBS website at http://www.pbs.org/wgbh/nova/manyworlds/orig-02.html.
13. Bryce S. DeWitt, quoted in Cécile DeWitt-Morette, *The Pursuit of Quantum Gravity: Memoirs of Bryce DeWitt from 1946 to 2004* (New York: Springer, 2011), p. 95.
14. Bryce S. DeWitt, phone interview by the author, December 4, 2002.
15. David Deutsch, quoted in Eugene Shikhovtsev, "Biographical Sketch of Hugh Everett III," https://space.mit.edu/home/tegmark/everett/everett.html.
16. Freeman Dyson, personal communication to the author, December 19, 2015.
17. Philip Ball, "Too Many Worlds," *Aeon Essays*, February 17, 2015, https://aeon.co/essays/is-the-many-worlds-hypothesis-just-a-fantasy.
18. Wojciech H. Zurek, "Decoherence and the Transition from Quantum to Classical," *Physics Today*, vol. 44, no. 10 (October 1991), p. 44.
19. David Wallace, "The Everett Interpretation," in Robert Batterman, editor, *Oxford Handbook of Philosophy of Physics* (Oxford: Oxford University Press, 2013), p. 471.
20. Simon Saunders, "Branch-Counting in the Everett Interpretation of Quantum Mechanics," *Proceedings of the Royal Society A*, vol. 477 (2021).

Chapter 4: Order from Chaos

1. David Kaiser, "How the Hippies Saved Physics: Science, Counterculture, and the Quantum Revival," *Scientific American*, January 30, 2012, https://www.scientificamerican.com/article/how-the-hippies-saved-physics-science-counterculture-and-quantum-revival-excerpt/.

2. Charles W. Misner, phone interview by the author, December 6, 2015.

3. John A. Wheeler, reported in Eugene Shikhovtsev, "Biographical Sketch of Hugh Everett III," https://space.mit.edu/home/tegmark/everett/everett.html#53.

4. Later supersymmetry theorists would reappropriate that term and bestow it with a different meaning.

5. Bryce S. DeWitt, phone interview by the author, December 4, 2002.

6. Charles W. Misner, interviewed by Alan Lightman, American Institute of Physics Niels Bohr Library and Archives Oral Histories, April 3, 1989, https://www.aip.org/history-programs/niels-bohr-library/oral-histories/33955.

7. P. James E. Peebles, "Robert Dicke and the Naissance of Experimental Gravity Physics, 1957–1967," *European Physical Journal H*, vol. 72 (2017), p. 186.

8. Robert H. Dicke, "The Principle of Equivalence and the Weak Interactions," *Reviews of Modern Physics*, vol. 29, no. 3 (July 1957), p. 357.

9. Robert H. Dicke and P. James E. Peebles, in S. W. Hawking and W. Israel, editors, *General Relativity: An Einstein Centenary Survey* (Cambridge, UK: Cambridge University Press, 1979), p. 504.

10. Robert H. Dicke, interviewed by Alan Lightman, American Institute of Physics Niels Bohr Library and Archives Oral Histories, January 19, 1988, https://www.aip.org/history-programs/niels-bohr-library/oral-histories/33931.

11. Roger Penrose, interviewed by Alan Lightman, American Institute of Physics Niels Bohr Library and Archives Oral Histories, January 24, 1989, https://www.aip.org/history-programs/niels-bohr-library/oral-histories/34322.

12. Bernard Carr, videoconference interview by the author, June 23, 2022.

13. Stanley Deser, "The Anthropic (and Mis-) Principle Revisited, Steven Weinberg in Memoriam," https://arxiv.org/abs/2202.09358.

Chapter 5: Burgeoning Truths

1. Dennis Overbye, "Space Ripples Reveal Big Bang's Smoking Gun," *New York Times*, March 18, 2014.

2. Andrei Linde, interviewed by David Zierler, American Institute of Physics Niels Bohr Library and Archives Oral Histories, January 26, 2021, https://www.aip.org/history-programs/niels-bohr-library/oral-histories/47185.

3. Paul Steinhardt, interviewed by David Zierler, American Institute of Physics Niels Bohr Library and Archives Oral Histories, June 2020, https://www.aip.org/history-programs/niels-bohr-library/oral-histories/46757.

4. Andrei Linde, interviewed by David Zierler, American Institute of Physics Niels Bohr Library and Archives Oral Histories, January 26, 2021, https://www.aip.org/history-programs/niels-bohr-library/oral-histories/47185.

5. Andrei Linde, interviewed by David Zierler, American Institute of Physics Niels Bohr Library and Archives Oral Histories, January 26, 2021, https://www.aip.org/history-programs/niels-bohr-library/oral-histories/47185.

6. Paul Steinhardt, interviewed by David Zierler, American Institute of Physics Niels Bohr Library and Archives Oral Histories, June 2020, https://www.aip.org/history-programs/niels-bohr-library/oral-histories/46757.

7. Jennifer Ouellette, "Multiverse Collisions May Dot the Sky," *Quanta Magazine*, November 12, 2014, https://www.quantamagazine.org/multiverse-collisions-may-dot-the-sky-20141110.

8. Hiranya Peiris, personal communication to the author, October 18, 2022.

9. John M. Kovac, quoted in Dennis Overbye, "Space Ripples Reveal Big Bang's Smoking Gun," *New York Times*, March 18, 2014.

10. Dennis Overbye, "Space Ripples Reveal Big Bang's Smoking Gun," *New York Times*, March 18, 2014.

11. Hiranya Peiris, personal communication to the author, October 18, 2022.

12. Hiranya Peiris, personal communication to the author, October 18, 2022.

Chapter 6: Tangled Up in Strings

1. Claud Lovelace, personal communication to the author, July 17, 2003.

2. Bernard Julia, interview by the author, Paris, January 13, 2003.

3. Bernard Julia, interview by the author, Paris, January 13, 2003.

4. Michael J. Duff, "A Layman's Guide to M-Theory," talk delivered at the Abdus Salam Memorial Meeting, International Centre for Theoretical Physics, Trieste, Italy, November 1997.

5. Michael R. Douglas and Shamit Kachru, "Flux Compactification," *Reviews of Modern Physics*, vol. 79, no. 2 (2007), p. 733.

6. Leonard Susskind, interviewed by David Zierler, American Institute of Physics Niels Bohr Library and Archives Oral Histories, May 1 and 3, 2020, https://www.aip.org/history-programs/niels-bohr-library/oral-histories/46752.

7. Raphael Bousso and Joseph Polchinski, "Quantization of Four-Form Fluxes and Dynamical Neutralization of the Cosmological Constant," *Journal of High Energy Physics*, vol. 2000 (July 2000).

8. Leonard Susskind, "The Anthropic Landscape of String Theory," reprinted in Bernard Carr, editor, *Universe or Multiverse?* (Cambridge, UK: Cambridge University Press, 2007), p. 248.

9. Peter Woit, "Weinberg Goes Anthropic," *Not Even Wrong* blog, November 4, 2005, https://www.math.columbia.edu/%7Ewoit/wordpress/?p=289.

10. Steven Weinberg, "Living in the Multiverse," reprinted in Bernard Carr, editor, *Universe or Multiverse?* (Cambridge, UK: Cambridge University Press, 2007), p. 30.

11. Steven Weinberg, interviewed by David Zierler, American Institute of Physics Niels Bohr Library and Archives Oral Histories, August 20, 2020, https://www.aip.org/history-programs/niels-bohr-library/oral-histories/47198.

12. Juan Maldacena, interviewed by David Zierler, American Institute of Physics Niels Bohr Library and Archives Oral Histories, January 15, 2021, https://www.aip.org/history-programs/niels-bohr-library/oral-histories/47184.

13. Gordon Kane, interviewed by David Zierler, American Institute of Physics Niels Bohr Library and Archives Oral Histories, April 4, 2021, https://www.aip.org/history-programs/niels-bohr-library/oral-histories/47087.

14. Paul Steinhardt, interviewed by David Zierler, American Institute of Physics Niels Bohr Library and Archives Oral Histories, June 2020, https://www.aip.org/history-programs/niels-bohr-library/oral-histories/46757.

Chapter 7: Seasons of Rebirth

1. Raman Sundrum, phone interview by the author, December 3, 2002.
2. Paul Steinhardt, interviewed by the author, September 13, 2002.
3. Paul Steinhardt, interviewed by the author, September 13, 2002.
4. Justin Khoury, personal communication to the author, September 13, 2011.
5. Paul Steinhardt, interviewed by the author, September 13, 2002.
6. Paul Steinhardt, interviewed by David Zierler, American Institute of Physics Niels Bohr Library and Archives Oral Histories, June 2020, https://www.aip.org/history-programs/niels-bohr-library/oral-histories/46757.
7. Stephen W. Hawking et al., quoted in Aric Jenkins, "Stephen Hawking and Fellow Scientists Dismiss 'Big Bounce' Theory in Letter," *Time*, May 13, 2017, https://time.com/4778304/stephen-hawking-scientific-american-letter-big-bounce/.
8. Justin Khoury, interviewed by the author, June 6, 2022.
9. Bernard Carr, videoconference interview by the author, June 23, 2022.
10. Roger Penrose, response to online question from the author, American Physical Society meeting (virtual), April 2021, https://phalpern.medium.com/dennis-sciamas-astonishing-cosmological-conversion-89473dd499d4.
11. "Cosmos May Show Echoes of Events Before Big Bang," *BBC News*, November 27, 2010.
12. Bartjan van Tent, Paola C. M. Delgado, and Ruth Durrer, "Constraining the Bispectrum from Bouncing Cosmologies with Planck," *Physical Review Letters*, vol. 130 (May 9, 2023), https://journals.aps.org/prl/abstract/10.1103/PhysRevLett.130.191002.
13. Albert Einstein to Vero and Bice Besso, March 21, 1955, quoted in Albrecht Fölsing, *Albert Einstein*, translated by Ewald Osers (New York: Penguin, 1997), p. 741.

Chapter 8: The Time Travelers Party

1. Larry Niven, "All the Myriad Ways," *Galaxy*, vol. 27, no. 3 (October 1968), p. 32.
2. Stephen W. Hawking, "Chronology Protection Conjecture," *Physical Review D*, vol. 46, no. 2 (July 15, 1992), p. 603.

NOTES

3. David Deutsch, *The Fabric of Reality: The Science of Parallel Universes and Its Implications* (New York: Penguin Books, 1997), p. 263.

Conclusion: The Reflecting Pool and the Sea

1. Martin Rees, "The Multiverse: Our Universe Is Suspiciously Unlikely to Exist—Unless It Is One of Many," *Singularity Hub*, April 9, 2023, https://singularityhub.com/2023/04/09/the-multiverse-our-universe-is-suspiciously-unlikely-to-exist-unless-it-is-one-of-many/.

2. William James, "Is Life Worth Living?" (1895), reproduced online by Monadnock Valley Press, https://monadnock.net/james/worth.html.

3. Ashley Fetters, "I Think About This a Lot: The Sliding Doors in *Sliding Doors*," *New York*, April 9, 2018, https://www.thecut.com/2018/04/i-think-about-this-a-lot-the-sliding-doors-in-sliding-doors.html.

4. Hannah Paine, "Diana's Sliding Doors Moment: How Princess' Life Was Ruled by Chance Decisions," News Corp Australia, September 1, 2018, https://www.news.com.au/entertainment/celebrity-life/royals/dianas-sliding-doors-moment-how-princess-life-was-ruled-by-chance-decisions/news-story/30bc740cd68c30d56651f0f1635d11f1.

5. Clarisse Loughrey, film review of *Doctor Strange in the Multiverse of Madness*, *The Independent*, May 3, 2022, https://www.independent.co.uk/arts-entertainment/films/reviews/doctor-strange-multiverse-of-madness-review-post-credit-scenes-spoilers-b2074031.html.

6. Paul Halpern and Nick Tomasello, "Size of the Observable Universe," *Advances in Astrophysics*, vol. 1, no. 3 (November 2016), https://dx.doi.org/10.22606/adap.2016.13001.

7. Paul Steinhardt, interviewed by David Zierler, American Institute of Physics Niels Bohr Library and Archives Oral Histories, June 2020, https://www.aip.org/history-programs/niels-bohr-library/oral-histories/46757.

8. George F. R. Ellis, "Does the Multiverse Really Exist?" *Scientific American*, August 1, 2011, https://www.scientificamerican.com/article/does-the-multiverse-really-exist/.

9. Virginia Trimble, personal communication to the author, May 13, 2022.

10. Philip Ball, "Experiments Spell Doom for Decades-Old Explanation of Quantum Weirdness," *Quanta Magazine*, November 2022, https://www.quantamagazine.org/physics-experiments-spell-doom-for-quantum-collapse-theory-20221020/.

INDEX

Abbott, Edwin A., 44
Abidor, Mitchell, 26
absolute space, 17, 48, 56–61
absolute time, 17, 48, 56–61
absolute zero, 132
abstract space, 6–18, 42–43, 54, 83–84, 96, 146
Adam and Eve, 33
ADD model, 221–223
aeon, 231–234, 275–276
Agullo, Ivan, 235
Albrecht, Andreas, 175
"All the Myriad Ways," 252
Almagest, 33
Alpher, Ralph, 139, 170
alternative realities, 9–18, 96–107, 127–133, 192–193, 220–228, 254–269
American Journal of Physics (journal), 247
amoeba, 80, 102, 107
Ampère, André-Marie, 49
An, Daniel, 234
Andersen, Hans Christian, 82
anisotropic cosmology, 128–131, 145–153
Anthropic Cosmological Principle, The (book), 158
anthropic cosmology, 4, 19–21
"Anthropic Landscape of String Theory, The," 211
Anthropic Principle
   Brandon Carter and, 19–21, 124–129, 153–163
   Charles Misner and, 124, 127–129, 133, 142–153, 158, 163–164
   concept of, 127–165
   explanation of, 4, 19–21
   Hugh Everett and, 158
   John Wheeler and, 129, 134, 148–151, 158
   Joseph Polchinski, 207, 210
   Leonard Susskind and, 209–213
   Raphael Bousso and, 210
   Roger Penrose and, 159–160
   Steven Weinberg and, 193, 208–214, 223
   string landscape multiverse and, 274
   Strong Anthropic Principle, 19, 124, 157–165, 193, 212
   Weak Anthropic Principle, 156, 158, 163
anthropic reasoning, 125, 158–168, 179, 209–214, 227
anti-de Sitter spaces, 222–223
antimatter, 150, 235–236, 272
antiparticles, 99, 235–238
antiquarks, 193–197
Aristarchus of Samos, 33–34
Aristotle, 30–31, 33–34
Arkani-Hamed, Nima, 221
Arnold, Harvey, 94
Arrhenius, Svante, 68–69
Ashok, Sujay, 207–208
Ashtekar, Abhay, 220, 234
Asner, Ed, 268
Aspect, Alain, 114
Aspen Center for Physics, 183
astral bodies, 17, 19, 28, 35, 134, 141
astral collisions, 184
astrophysics, 54
atomic clocks, 243
atomic components, 35–36, 88–89, 195
atomic era, 93
atomic models, 69–70
atomic scale, 120, 151–152
atomic time, 151–152, 243

Back to the Future (film), 266
Background Imaging of Cosmic Extragalactic Polarization 2 (BICEP 2), 186
Bahnson, Agnew H. Jr., 105

INDEX

Bahnson Company, 105
Ball, Philip, 113
Bargmann, Valentine "Valya," 73
Barish, Barry, 246
Barrow, John, 127, 149, 158, 178
baryons, 193, 197
Battle of Waterloo, 36
BBC News, 233
Belinsky, Vladimir, 147–148
Bell, John Stewart, 114
Bergmann, Peter, 73, 104, 195, 197, 199, 203
Bewitched (TV series), 268
Beyer, Anna, 65
Bianchi, Luigi, 147–148
Bianchi Type IX, 148, 152
Bianchi types, 147–148, 152, 164
BICEP 2 (Background Imaging of Cosmic Extragalactic Polarization 2), 186
Big Bang theory
  cosmic strings and, 249
  cyclic cosmologies and, 219–236
  explanation of, 1, 17, 20, 46, 131–143, 147–160, 164–170, 219–236, 274
  George Gamow and, 139, 170, 246
  Jim Peebles and, 139, 153
  naming of, 131
  particle physics and, 164
  Robert Dicke and, 153
  Roger Penrose and, 160
  Stephen Hawking and, 138–139, 152–153, 176, 184
Big Bounce theories, 217, 220, 228–230, 234–236, 275
Big Crunch model, 153, 219–220
BKL team, 147–148
black box, 87–88, 96, 115
Black Hole, The (film), 269
black holes, 65, 104, 134–139, 153–155, 231–234, 245–246, 253, 269–274
Blanqui, Louis-Auguste, 26–27, 32–37, 46, 57, 159, 161, 256
Bohm, David, 114, 191
Bohr, Niels
  Albert Einstein and, 24, 77, 87–90, 96
  black box approach, 115
  complementarity and, 87–88, 115
  death of, 110
  decoherence theory and, 115–116
  Hugh Everett and, 13, 79, 97–104
  John Wheeler and, 97–104, 109–110

  as Nobel laureate, 87
  Oskar Klein and, 68–70
  philosophy of, 6–8
  photos of, *87*, *91*
  quantum visions of, 81–82, 228
Bojowald, Martin, 220, 234
Boltzmann, Ludwig, 56, 120, 159
Boltzmann brains, 160–161
Born, Max, 74, 85, 117–120
Born rule, 117–120
bosons, 169–171, 178, 188, 196–204, 216, 222, 273
bouncing universes, 217, 220, 228–230, 234–236, 275
Bousso, Raphael, 210
Boyle, Latham, 220, 235–236
Bradbury, Ray, 251
Brahe, Tycho, 34
branching realities, 76–77, 107–113, 120–121, 250–256
branching timelines, 238, 243, 250–256, 275
brane worlds, 60, 206–207, 217, 220–227, 230, 237
Brans, Carl, 133, 151
Brans-Dicke theory, 133, 151, 203
Breakthrough Prize in Fundamental Physics, 111, 184
Brief History of Time, A (book), 184
Brill, Dieter, 105, 146
Bring the Jubilee (book), 252
Bruno, Giordano, 25, 34
bubble collisions, 21–22, 183–187, 227–230
bubble universes
  collisions and, 21–22, 183–187, 227–230
  eternal inflation and, 163–164, 183–193, 210–218, 236–237, 265–267, 270–276
  false vacuum bubbles, 173–175
  inflationary bubbles, 20–24, 163–164, 173–176, 183–188, 227–230, 275–276
  notion of, 20–24, 163–164, 180–183
  scalar field and, 174–175
Burnell, Jocelyn Bell, 137
butterfly effect, 149

Caius College, 240
Calabi, Eugenio, 73, 205
Calabi-Yau spaces, 73, 205–211, 217, 237
Caltech, 162, 198, 246
Candelas, Philip, 205
Candide (book), 16

INDEX

Capra, Frank, 259
cards, house of, 7, 87, 100
Carr, Bernard, 161–162, 213, 230
Carter, Brandon
   Anthropic Principle and, 19–21, 124–129, 153–163
   birth of, 110
   black holes and, 153–155
   early years of, 110
   John Wheeler and, 155–158
   "Large Number Coincidences and the Anthropic Principle in Cosmology" and, 126, 156, 163
   multiple universes and, 153–158
   Strong Anthropic Principle and, 19, 124, 157–159, 163
cat paradox, 7–9, 85, 107, 115, 118
Catholic University of America, 82, 92
cave analogy, 75
Central Fire, 29, 33
CERN, 194
Chao-Lin Kuo, 186
chaos theory, 47
chaotic behavior, 47, 127–166, 173–176, 181–182, 265
chaotic cosmology programme, 127, 149–150
chaotic inflation, 181–182
Chaucer, Geoffrey, 64
Christian beliefs, 33
chromodynamics, 197, 200–201
"Chronology Protection Conjecture," 238, 253–254
Civil War, 252
classical mechanics, 5, 38, 56, 94
classical physics, 6, 18, 30–31
Clauser, John, 114
Clausius, Rudolf, 44–45
Clerke, Agnes Mary, 272
Closed Timelike Curves (CTCs), 244, 248–251, 254
closed trapped surfaces, 138
CMBR (Cosmic Microwave Background Radiation), 132–151, 163–168, 174–188, 220–236, 260, 274–275
COBE (Cosmic Background Explorer), 141, 178
Cold War, 93, 103
Collins, Christopher Barry, 152–156
color charge, 197
Columbia University, 171, 212

compactification, 3, 73, 199–217
complementarity philosophy, 87–88, 115
Conformal Cyclic Cosmology, 24, 217, 220, 231–237
conservation laws, 38–46, 52–53, 148, 272
conservation of energy, 39–41, 44, 52–53
conservative forces, 39–40
Contact (book), 245
Copenhagen interpretation, 6–7, 15, 83–89, 97–107, 113–115, 120–123
Copernican Principle, 155
Copernicus, Nicolaus, 33–34, 102, 108, 155
Cornell University, 171, 212, 258
"corpuscles," 31, 201
Cosmic Background Explorer (COBE), 141, 178
Cosmic Microwave Background Radiation (CMBR), 132–151, 163–168, 174–188, 220–236, 260, 274–275
cosmological constant, 17–19, 128–131, 140, 168–188, 190–217, 222–224, 260
cosmology
   anisotropic cosmology, 128–131, 145–153
   anthropic cosmology, 4, 19–21, 158
   chaotic cosmology, 127, 149–150
   Conformal Cyclic Cosmology, 24, 217, 220, 231–237
   cyclic cosmologies, 23–24, 217–239, 241
   explanation of, 4, 11–21, 53
   isotropic cosmology, 131, 150
   loop quantum cosmology, 220, 234–236
   Pythagorean cosmology, 29
   standard cosmology, 18–20
   vacuum cosmology, 148
Cosmos (book), 184
"Creation of Universes from Nothing," 180
Cremmer, Eugène, 199–201, 203
Cronenberg, David, 267
Crookes, William, 53
CTCs (Closed Timelike Curves), 244, 248–251, 254
cyclic cosmologies
   Big Bang and, 219–236
   concept of, 23–24, 219–239, 241
   Conformal Cyclic Cosmology, 24, 217, 220, 231–237
   explanation of, 24, 217, 219–220
   Paul Steinhardt and, 23–24, 219–220, 224–233, 237
   rebirth and, 219–239

cyclic cosmologies (*continued*)
  Roger Penrose and, 23–24, 217–220, 230–236
Cyclic Universe, 23–24, 220, 225–228, 236
cyclic universe models, 23–24, 217, 219–239, 241
cylinder condition, 66–67, 199, 244–245

Dalton, John, 30–31, 35
damping forces, 39–40
dark energy, 140, 191, 209, 227, 233–236, 259, 274
dark matter, 140, 191, 233–236, 259, 274
Darwin, Charles, 254
Darwinism, 53, 136
DC Comics, 269
de Broglie, Louis, 69–71, 76–77, 85, 114
de Camp, L. Sprague, 251
de Sitter, Willem, 17–18, 130–131, 168–170, 173, 222–223
decoherence theory, 114–116, 120–122
Delgado, Paola C. M., 235
Democritus, 30–31
Deser, Stanley, 162–163, 201
Deutsch, David, 111, 119, 238, 254–255
DeWitt, Bryce, 9–10, 104–112, *105*, 127, 154, 197
DeWitt-Morette, Cécile, 104, 110
Diana, Princess, 265–266
"dice-rolling," 73–76, 96
Dick, Philip K., 252
Dicke, Robert H. "Bob"
  Big Bang theory and, 153
  Brans-Dicke theory and, 133, 151, 203
  CMBR and, 132–134, 139, 141
  flatness problem and, 153, 171
  general relativity and, 124, 139, 151
  gravitational theory and, 151
  isotropic universe and, 152–153
  Jim Peebles and, 132, 151–153
  Large Numbers Hypothesis and, 151–153, 155–156
  multiple universes and, 153–158
  photo of, *151*
Die fröhliche Wissenschaft (*The Joyful Science*) (book), 37
dimensions
  concept of, 43–44

doughnut dimensions, 68–73
eleven-dimensional space, 14–15, 60, 72, 193, 201, 203, 206
examples of, *43*, 43–44
fifth dimension, 13–15, 23, 60, 65–76, 84, 195, 199, 203
five-dimensional unification, 13–15, 60–76, 195, 199, 203
four-dimensional space, 12–14, 42–44, 60–66, 75, 164, 191, 201, 204
fourth dimension, 5, 11–12, 15, *43*, 54–57, 63–66, 70, 75, 84, 191–192, 195
higher-dimensional spaces, 3, 11–14, 42–43, 55–61, 68, 72–74, 164, 188, 191–217, 220–228
quantum entanglement and, 191–217
six-dimensional space, 192, 195–198, 204–205
spatial dimensions, 54–55, 147, 207
strings and, 188, 191–217
ten-dimensional theory, 14, 198–200, 204–206
three dimensions, 11, 14–17, *43*, 43–44, 54–55, 61, 72–73, 128, 221
three-dimensional objects, *43*, 44, 54, 72, 145–146
three-dimensional space, 11, 14, 72–73, 221
two dimensions, *43*, 44, 75, 206
Dimopoulos, Savas, 221
Diósi, Lajos, 122
Dirac, Paul
  general relativity and, 202
  gravitational theory and, 150–151, 202
  Large Number Hypothesis and, 150–151, 176–177
  as Nobel laureate, 132, 150
Dirichlet brane, 207, 221
Doctor Strange in the Multiverse of Madness (film), 10, 269
doppelgängers, 11, 256, 261–264
doughnut shapes, 68–73, 200, 205
Douglas, Michael R., 207
Doyle, Arthur Conan, 53
Dresher, Melvin, 93
dual resonance theory, 194
Duff, Michael, 206
Durrer, Ruth, 235
Dvali, Gia, 221
Dyson, Freeman, 112–113

INDEX

École Normale Supérieure, 199
Eddington, Arthur, 132, 150, 266
eigenstates, 86–88, 117
Einstein, Albert
  birthday of, 202, 240, 242
  cosmological constant and, 208
  death of, 73, 128, 195
  "dice-rolling" and, 73–76
  early years of, 52
  Einstein-Bergmann model, 203
  Einstein-de Sitter universe, 17–18, 130–131
  electromagnetism and, 13, 67
  expanding universe and, 17–18, 208
  final lecture by, 97
  flow of time and, 238–239
  general relativity and, 11–13, 16–18, 58–68, 74–76, 84–85, 104–106, 128–129, 138, 145–149, 162, 169, 191, 202, 237–238, 241–244
  gift for, 240, 242
  gravitational theory and, 149–150, 159
  higher-dimensional spaces and, 199
  Hugh Everett and, 92
  Institute for Advanced Study and, 83
  John Wheeler and, 77–78
  later years of, 73–75, 104
  mouse and, 7, 76–78, 97, 99
  Niels Bohr and, 24, 77, 87–90, 96
  as Nobel laureate, 61
  photo of, *61*
  "poker-playing" and, 73–76
  quantum entanglement and, 114
  quantum measurement and, 7, 76–78, 97–99
  special relativity and, 11–12, 61–65, 195, 237–238, 243
  Theodor Kaluza and, 58–60, 64–68
Einstein-Bergmann model, 203
Einstein-de Sitter universe, 17–18, 130–131
Einstein's mouse, 7, 76–78, 97, 99
Ekpyrotic Universe, 23–24, 220, 224–229
electromagnetism
  Alan Guth and, 168–171
  Albert Einstein and, 13, 67
  electricity and, 48–50, 59–60, 273
  electromagnetic radiation, 50–51
  explanation of, 3, 13
  geons and, 146, 246
  gravitation and, 13–15, 60, 64–72, 94, 146, 195–196, 221
  James Clerk Maxwell and, 48–51, 56, 59, 64, 67, 94, 273
  Kaluza-Klein theory and, 13, 72–73, 195
  model of, 171
  Oskar Klein and, 13, 69–73
  quantum electrodynamics and, 94
  radiation and, 50–51
  range of, 19
  Steven Weinberg and, 171
  Theodor Kaluza and, 13, 60, 64–67, 72–73
  theory of, 49–51, 56, 59–60, 70, 94–95
electrons, 7, 67–77, 85–120, 139–140, 196–207, 235–236
electroweak theory, 73, 164, 171, 193, 197, 200–201, 212, 273
electroweak unification, 171, 197, 212, 273
eleven-dimensional space, 14–15, 60, 72, 193, 201, 203, 206
Elizabeth II, Queen, 81
Ellis, George, 270–272
Empedocles, 29–30
endless loops, 38–44
energy
  conservation of, 39–41, 44, 52–53
  dark energy, 140, 191, 209, 227, 233–236, 259, 274
  entropy and, 44–45
  kinetic energy, 39–40, 120
  mechanical energy, 39–41, 44, 52–53, 94–95, 116
  of motion, 39–40
  of position, 39–40
  potential energy, 39–40, 175–181
entropy, 44–45, 219–220
epicycles, 34
epochs, 24, 139, 148–149, 171, 177–181
eternal inflation
  bubble universes and, 163–164, 183–193, 210–218, 236–237, 265–267, 270–276
  expanding universe and, 21–24, 181–182, 270–271
  explanation of, 21–24, 163–164
  inflationary universe and, 163–164, 179–193, 210–218, 224–237, 275–276
  Paul Steinhardt and, 21–24, 179–183, 216–217, 224–230, 270–272
eternal recurrence, 36
eternal return, 23, 36–46, 134
eternity, 26–57, 166, 227
Eternity Through the Stars (book), 26, 32

INDEX

ETH (Swiss Federal Institute of Technology), 61–62
ether, 51–57, 60
ether wind, 51
event horizon, 135–138, 272–274. *See also* horizon problems
Everett, Hugh
   Albert Einstein and, 92
   Anthropic Principle and, 158
   birth of, 91
   Charles Misner and, 78, 94, 97–105
   daughter of, 112, 253
   death of, 112
   documentary about, 112
   early years of, 82, 90–94
   John Wheeler and, 8–9, 97–104, 106–110, 144
   Many Worlds Interpretation and, 76, 91, 105–116, 121–124, 157–158, 188, 192, 214, 238, 254, 263
   Niels Bohr and, 13, 79, 97–104
   photo of, *91*
   quantum measurement and, 8–9, 76–81, 93–102, 111–123, 192
Everett, Hugh Jr., 91
Everett, Katherine Kennedy, 91–92
Everett, Liz, 112, 253
Everett, Mark, 112
Everett, Nancy, 112
Everything Everywhere All at Once (film), 2, 11, 261, 263
expanding universe
   Albert Einstein and, 17–18, 208
   bouncing universes and, 217, 220, 228–230, 234–236, 275
   eternal inflation and, 21–24, 181–182, 270–271
   evidence of, 17, 130–131, 185–189
   Hubble telescope and, 130–131
   rate of, 270

Fabric of Reality, The (book), 254
false vacuum bubbles, 173–175
Faraday, Michael, 49
Fayed, Dodi, 266
fermions, 196–204, 216
Ferrara, Sergio, 201
Fetters, Ashley, 265
Feynman, Richard
   antiparticles and, 236

   gravitation and, 104
   multiverse theory and, 258
   string theory and, 216
   "sum over histories" and, 6, 18, 94–101, 146
field theory, 13, 49–50, 56
fifth dimension, 13–15, 23, 60, 65–76, 84, 195, 199, 203
Finn, Kieran, 220, 235–236
five-dimensional tensors, 65–66
five-dimensional unification, 13–15, 23, 60–76, 195, 199, 203
Flash, The (film), 269
Flatland (book), 44
flatness problem, 151–153, 163, 168–176, 274
Flood, Merrill, 93
FLRW metrics, 131–132, 142–143, 147–152, 156, 174–176
"Fluctuations in the New Inflationary Universe," 178
force carriers, 14, 171, 196–198
Ford, Kenneth, 89
four-dimensional space, 12–14, 42–44, 60–66, 75, 164, 191, 201, 204
four-dimensional tensors, 65–66
four-dimensional thinking, 60–66
fourth dimension, 5, 11–12, 15, *43*, 54–57, 63–66, 70, 75, 84, 191–192, 195
Fragments of Science (book), 41, 52–53, 272
frame-dragging, 244–245
Freedman, Daniel, 201
Friedmann, Alexander, 131–132
Friedmann-Lemaître-Robertson-Walker (FLRW) metrics, 131–132, 142–143, 147–152, 156, 174–176
fusion, cycles of, 140
fusion, latent heat of, 172–173

galaxies
   distant galaxies, 1, 130, 142–143, 154, 208
   Milky Way, 25, 155–157, 174, 233, 270, 272
   number of, 269–270
Galaxy (magazine), 252
Galilean relativity, 51–52, 56
Galileo, 25, 35, 51–52
Gamow, George, 84, 139, 170, 246
Garden of Eden, 154
Gauss, Carl, 43, 49

INDEX

Gell-Mann, Murray, 150, 204
general relativity
   Albert Einstein and, 11–13, 16–18, 58–68, 74–76, 84–85, 104–106, 128–129, 145–149, 162, 169, 191, 202, 237–238, 241–244
   closed trapped surfaces and, 138
   Edward Witten and, 201–202
   explanation of, 11–13, 237
   John Wheeler and, 77–78, 94–99, 104–105
   Louis Witten and, 201–202
   meeting on, 202
   Paul Dirac and, 202
   quantum theory and, 99, 241–242
   Robert Dicke and, 124, 139, 151
   Roger Penrose and, 138
   Stephen Hawking and, 238
   tensors of, 65–66
   see also special relativity
geocentric beliefs, 33–34, 102
geometric foam, 18. *See also* quantum foam
geons, 146, 246
Ghirardi, Giancarlo, 122
Glashow, Sheldon, 171, 216, 273
gluons, 171, 194, 197, 200
Gödel, Kurt, 240, 242–245
Goldman, Alan, *225*
googol concept, 147, 193
Gott, J. Richard, 249
Goudsmit, Samuel, 69
graceful exit problem, 175–176
Graduate College, Princeton University, 94, 97–98
grand unification theories (GUTs), 171–174, 232
Grandfather Paradox, 250–251
gravitation
   electromagnetism and, 13–15, 60, 64–72, 94, 146, 195–196, 221
   hierarchy problem, 221
   loop quantum gravity, 234, 276
   quantum theory of, 18, 108–109, 145, 170, 197–200, 234, 275
   supergravity, 14, 72, 164, 201–208
Gravitation (book), 246
gravitational theory
   Albert Einstein and, 149–150, 159
   Charles Misner and, 142
   Edward Witten and, 201–202
   John Wheeler and, 18–19, 104

   Kip Thorne and, 237
   Newtonian gravitation and, 47–48, 63
   Paul Dirac and, 150–151, 202
   Robert Dicke and, 151
   Roger Penrose and, 159
gravitational waves, 105, 170, 186, 230, 245–246, 275
gravitons, 198, 201, 204, 207, 221–224
Greek philosophers, 25, 28–33
Green, Michael, 203–204
Groundhog Day (film), 14
Gurzadyan, Vahe G., 233
Guth, Alan
   birth of, 170
   electromagnetism and, 168–171
   flatness problem and, 168–176
   inflationary universe and, 20–21, 166–178, 186–187, 224
GUTs (grand unification theories), 171–174, 232

hadrodynamics, 194
hadronic string theory, 194–197. *See also* string theory
hadrons, 193–194
harmonic oscillator potential, 181
Harrison, David, *91*
Harvard University, 186, 205, 212–213
Harvard-Smithsonian Center for Astrophysics, 186
Hawking, Stephen
   as author, 184
   Big Bang theory and, 138–139, 152–153, 176, 184
   "Chronology Protection Conjecture" by, 238, 253–254
   entropy and, 219–220
   general relativity and, 238
   inflationary universe and, 229
   isotropic universe and, 152–154
   Kip Thorne and, 253–254
   multiple universes and, 153–156
   photo of, *138*
   Roger Penrose and, 138, 184
   scalar field and, 178
   singularity and, 128
   time travel and, 240–241, 253–254
Hawking points, 234
Hawking radiation, 135, 232, 234, 272
Heaviside, Oliver, 59

INDEX

Heisenberg, Werner, 5, 74, 88–89, *89*, 95–96, 106
Heisenberg's uncertainty principle, 5, 88–89, 95–96
Henry II, King, 267
Herman, Robert, 139
Hertz, Heinrich, 273
hierarchy problem, 221–223
"Hierarchy Problem and New Dimensions at a Millimeter, The," 221
Higgs, Peter, 169
Higgs boson, 169–171, 178, 188, 222, 273
Higgs mechanism, 169–171
Higgs scalar field, 169–171, 178
higher-dimensional spaces
  concept of, 3, 11–14, 55–61, 68, 73–74, 164, 188
  Edward Witten and, 191, 201–206, 211
  examples of, 42–44, *43*
  Steven Weinberg and, 191, 193, 204, 208–214
  string landscape multiverse and, 191–217
  string theory and, 191–217, 220–228
  superstring theory and, 164–165, 198–208
  see also dimensions
higher-dimensional unification, 3, 13, 60, 68, 195, 199, 202
Hilbert, David, 84
Hilbert space, 6–7, 15, 81–125, 146, 191–192, 274
Hilbert's hotel analogy, 84, 122–123. *See also* Hilbert space
Hinton, Charles, 44
Hitler, Adolf, 251
Hooker telescope, 141
Horava, Petr, 206
Horgan, John, 22, 260
horizon problem, 135–148, 163, 174, 176, 227, 272–274
"How the Leopard Got Its Spots," 190
Howitt, Peter, 265–266
Hoyle, Fred, 131, 176
Hubble, Edwin, 130–131, 141–142, 208
Hubble crisis, 270
Hubble expansion, 168–170, 174–175, 208–209
Hubble space telescope, 130–131, 141, 177
Hubble's law, 142–143, 208
Huggins, Margaret, 35
Huggins, William, 35

human observers, 7–8, 77–78, 84–90
hydrogen bomb project, 105
hypercube, *43*, 44, 54, 56
hyperspace, 23, 52–57, 59
hyperspheres, 44–45, 54

IAS (Institute for Advanced Study), 83, 93, 97, 214
Ijjas, Anna, 22, 228–229
inertia, 30–31, 47–48, 250
inflationary bubbles, 20–24, 163–164, 173–176, 183–188, 227–230, 275–276
inflationary universe
  Alan Guth and, 20–21, 166–178, 186–187, 224
  Alexei Starobinsky and, 170–171, 224
  Big Bounce theories and, 220, 228–230, 234–236, 275
  eternal inflation and, 163–164, 179–193, 210–218, 224–237, 275–276
  explanation of, 20–24, 163–164, 166–190
  graceful exit problem, 175–176
  new inflation, 175–181, 225
  Paul Steinhardt and, 20, 22–24, 224–230
  Stephen Hawking and, 229
  string theory and, 192–193, 210–217, 224–234
  see also bubble universes
"Inflationary Universe: A Possible Solution to the Horizon and Flatness Problems," 168
Institute for Advanced Study (IAS), 83, 93, 97, 214
Institute for Theoretical Physics, 87, 212
Institute of Field Physics, 104–105
intelligent life, 4, 16–19, 129, 153–155, 161–163, 178, 193
Interstellar (film), 12, 247
invariants, 12, 178, 187–188, 224, 229
"Is Life Worth Living?," 261
isotropic cosmology, 131, 150
isotropic expansion, 18, 231–232
isotropic universe, 128–132, 149–158, 175, 231–232
It's a Wonderful Life (film), 259

James, William, 10, 261–262
James Webb Space Telescope (JWST), 1–2, 25, 141, 177
Jammer, Max, 109
Johnson, Matthew, 21, 183–185

INDEX

Jordan, Pascual, 133
Jordan-Brans-Dicke theory, 133
Journal of Experimental and Theoretical Physics (journal), 170
Jow, Dylan, 234
Julia, Bernard, 201–203
Jupiter, 29, 33, 48
Just So Stories (book), 190

Kachru, Shamit, 208
Kaiser, David, 127
Kallosh, Renata, 186
Kaluza, Anna Beyer, 65
Kaluza, Dorothea, 65
Kaluza, Max, 64
Kaluza, Theodor
    Albert Einstein and, 58–60, 64–68
    birth of, 64
    early years of, 64–65
    electromagnetism and, 13, 60, 64–67, 72–73
    Kaluza-Klein theories, 13, 15, 68, 72–73, 84, 127, 188, 192, 195, 199–203
Kaluza, Theodor Jr., 65
Kaluza-Klein theories, 13, 15, 68, 72–73, 84, 127, 188, 192, 195, 199–203
Kane, Gordon, 214
Kasner, Edward, 147–149, 193
Kavli Institute for Theoretical Physics (KITP), 111
Kelly, Walt, 90
Kennedy, Katherine, 91–92
Kepler, Johannes, 34
Kerr, Roy, 136–137
Kerr-Newman solution, 137
Khalatnikov, Isaak, 147–148
Khoury, Justin, 23, 220, 226, 229–230
kinetic energy, 39–40, 120
Kipling, Rudyard, 190
Klein, Antonie "Toni" Levy, 68
Klein, Felix, 59
Klein, Gottlieb, 68
Klein, Oskar
    birth of, 68
    doughnut dimensions and, 68–73
    early years of, 68–69
    electromagnetism and, 13, 69–73
    Kaluza-Klein theories, 13, 15, 68, 72–73, 84, 127, 188, 192, 195, 199–203
    Niels Bohr and, 68–70

Klein bottles, 59
Kovac, John M., 186
Kuo, Chao-Lin, 186
Kurt, Gödel, 83

Laflamme, Raymond, 220
Landau Institute for Theoretical Physics, 147, 170
Laplace, Pierre-Simon, 30–31, 38
Large Hadron Collider, 169, 216, 222–224, 275
"Large Number Coincidences and the Anthropic Principle in Cosmology," 126, 156, 163
Large Number Hypothesis (LNH), 150–156, 176–177
Laser Interferometer Gravitational-Wave Observatory (LIGO), 246
latent heat of fusion, 172–173
Laub, Jakob, 62
laws, conservation, 38–46, 52–53, 148, 272
laws of mechanics, 94–95, 116
laws of motion, 30–31, 38–40, 48, 52, 192
laws of physics, 23, 124, 238, 249, 254–255
laws of thermodynamics, 44–46, 87, 115, 160
Leathes, Margaret, 231
Leibniz, Gottfried, 16, 82, 155
Leiden University, 103
Lemaître, Georges, 131–132
Lennon, John, 259
Leonardo, 119
leptons, 164, 188, 193
Lest Darkness Fall (book), 251
Lifshitz, Evgeny, 147–148
light waves, 51–52
light-years, 243, 270
LIGO (Laser Interferometer Gravitational-Wave Observatory), 246
Linde, Andrei, 20, 24, 167, 175–182, 186–187, 211, 224, 229, 260
linear momentum, 38–39, 42, 70
Lion in Winter, The (film), 267
LiteBIRD satellite, 187
"Living in the Multiverse," 213
Livio, Mario, 214
LNH (Large Number Hypothesis), 150–156, 176–177
Loeb, Abraham "Avi," 22, 228–229
Loki (TV series), 262
loop quantum cosmology, 220, 234–236

INDEX

loop quantum gravity, 234, 276
loops, endless, 38–44
Lorenz, Edward, 149
lost horizon, 139–145. *See also* horizon problems
Lou Grant (TV series), 268
Loughrey, Clarisse, 268
Lovelace, Claud, 195, 198
Low, Francis, 170
luminiferous ether, 51–57, 60

Mach, Ernst, 48, 56
magnetism, 48–50, 59–60, 221, 239, 273. *See also* electromagnetism
Maldecena, Juan, 214
Man in the High Castle, The (book), 252
Man in the High Castle, The (TV series), 2
manifold concept, 73, 193. *See also* compactification
"Many Minds" interpretation, 117, 121
Many Worlds Interpretation (MWI)
   Bryce DeWitt and, 9–10, 104–112, 127, 154
   explanation of, 10–11, 21–22
   Hugh Everett and, 76, 91, 105–116, 121–124, 157–158, 188, 192, 214, 238, 254, 263
   Larry Niven and, 252–253
   myths about, 261–270
   Paul Steinhardt and, 166, 258
   public view of, 256–257
   see also multiverse
"Many-Universes Interpretation of Quantum Mechanics," 9–10, 110
Marcel Grossmann Meeting on General Relativity, 202
Mars, 22, 29, 41
Marvel Cinematic Universe (MCU), 2, 10, 261–262, 267–269
Mary Tyler Moore Show, The (TV series), 268
matter, dark, 140, 191, 233–236, 259, 274
matter waves theory, 69–71, 85
Maxwell, James Clerk, 47–52, 55–56, 59–60, 64–67, 94, 106, 273
Maxwell's equations, 60, 64–67, 94
Maxwell's theory, 49–52, 55–56, 94, 273
Maxwell's unification, 49, 60, 273
Measuring and Modeling the Universe (book), 166
mechanical energy
   conservation of, 39–41, 44, 52–53
   kinetic energy and, 39–40
   laws of mechanics and, 94–95, 116
   potential energy and, 39–40
   see also energy
Meissner, Krzysztof, 234
Mendeleev, Dmitri, 35
Mercury, 25, 29
mesons, 193
Michelson, Albert, 51–52
Michelson-Morley experiment, 51–52
Milky Way, 25, 155, 157, 174, 233, 270, 272
Minkowski, Hermann, 12, 15, 61–63, *62*, 65, 84
"Mirror, Mirror," 261, 264
mirrors, house of, 99–100
Misner, Charles
   Anthropic Principle and, 124, 127–129, 133, 142–153, 158, 163–164
   as author, 246
   death of, 94
   early years of, 78, 94
   gravitational theory and, 142
   Hugh Everett and, 78, 94, 97–105
   John Wheeler and, 78, 94, 101–104
   Mixmaster universe and, 123–124, 127–165
   photos of, *91*, *142*
   quantum electrodynamics and, 94, 97
   quantum measurement and, 97–102
MIT, 11, 168, 170–171, 212
Mixmaster universe, 123–124, 127–165
momentum, linear, 38–39, 42, 70
Mona Lisa painting, 119
Moorcock, Michael, 10, 261–262
Moore, Ward, 252
moral multiverse, 261–265
Morley, Edward, 51–52
Morris, Michael, 246–251
motion, laws of, 30–31, 38–40, 48, 52, 192
Mount Wilson Observatory, 141
M-theory
   cosmological constant and, 208–210, 217
   Edward Witten and, 202, 206
   explanation of, 14
   Paul Steinhardt and, 217, 226
   string theory and, 14, 60, 72, 193, 202–210, 215–217, 221, 226–228, 234–237, 273–276

INDEX

superstring theory and, 72–73, 206, 215–217
multiverse
  concept of, 1–25, 27–57, 59–79, 81–125, 127–166, 176–276
  entertainment and, 1, 10–11, 261–270
  explanation of, 1–25
  fictional adventures, 261–264
  meaning of, 259–276
  moral multiverse, 261–265
  myths of, 261–270
  notion of, 1–25, 27–57, 59–79, 81–125, 127–276
  public view of, 256–257
  purpose of, 259–276
  reflecting on, 259–276
  string landscape multiverse, 14–21, 60, 188–189, 191–217, 223, 270–276
MWI. *See* Many Worlds Interpretation

Nambu, Yoichiro, 194, 197
Napoleon, 36
neutrinos, 236, 244
neutrons, 136, 139, 150, 169, 193, 232
Neveu, André, 198–199
new inflation, 175–181, 225
New Inflationary Universe, 178
New York Times (newspaper), 186
Newman, Ezra "Ted," 137
Newton, Isaac, 4–6, *30*, 30–31, 35, 38, 82, 106, 201
Newtonian gravitation, 47–48, 63
Newtonian mechanics, 5, 31, 38, 46–49, 56, 89, 94, 116, 119
Newtonian physics, 5–6, 17–18, 30–31, 41, 46–48, 53, 60, 63, 74, 89, 108, 192
Nielsen, Holger, 194–195
Nietzsche, Friedrich, 23, 27, 36–37, 42–46, 57, 134, 159, 161, 256
Niven, Larry, 245, 252–253
Nobel Institute, 68
Nobel laureates
  Adam Riess, 209
  Alain Aspect, 114
  Albert Einstein, 61
  Anton Zeilinger, 114
  Barry Barish, 246
  Brian Schmidt, 209
  Daniel Shechtman, 225
  Erwin Schrödinger, 85
  John Clauser, 114
  Kip Thorne, 246
  Murray Gell-Mann, 150, 204
  Niels Bohr, 87
  Paul Dirac, 132, 150
  Rainer Weiss, 246
  Roger Penrose, 138, 231
  Saul Perlmutter, 209
  Steven Weinberg, 188, 193, 204, 212
  Svante Arrhenius, 68
  Werner Heisenberg, 89
Noether, Emmy, 38–39
Nolan, Christopher, 12
Nordström, Gunnar, 68, 136–137
Novikov, Igor, 179, 251
"Novikov Self-Consistency Principle," 251
nuclear forces, 3, 41, 60, 96, 137, 153–157, 193
nuclear war, 93, 103–104
Nuffield Workshop, 176–177
numerical schemes, 11, 149–156
Nurowski, Pawel, 234

Occam's razor, 112
Ockham, William of, 112
omega parameter, 153
"On the Foundations of Quantum Mechanics," 80
On the Infinite Universe and Worlds (book), 34
On the Revolutions of the Heavenly Orbs (book), 33–34, 102
"On the Role of Gravitation in Physics," 104
On the Theory of Parlor Games (paper), 93
"On the Three-Body Problem and the Equations of Dynamics," 44
"On the Unity Problem of Physics," 66–67
one dimension, *43*, 61, 71
optical illusions, 54
orbits, 33–34, 41
Oscillation Project with Emulsion-tRacking Apparatus (OPERA), 244
oscillatory models, 133–134, 148, 181, 219, 244
"Outline of Feynman Quantization of General Relativity; Derivation of Field Equations; Vanishing of the Hamiltonian," 97
Ovrut, Burt, 23, 220, 226

INDEX

Page, Don, 46
parabola, 95, 181
"Paradoxical Ode, A," 56
parallel realities, 82, 121, 130, 252, 257–267
parallel universes, 2–13, 23–27, 36, 79, 112, 124, 154–164, 257–269
Parallel Worlds, Parallel Lives (documentary), 112
paranormal, 53–56
Paris Commune, 32
Participatory Universe idea, 110–112
particle physics
   antiparticles and, 99, 235–238
   Big Bang theory and, 164
   exploring, 99, 104
   predictions in, 155
   Standard Model of, 120, 164, 171–172, 192–193, 201–208, 212, 222–223, 226
   unification mechanisms, 169–173
Paul III, Pope, 102
Pauli, Wolfgang, 8, 13, 72, 74–75, 113
Pearle, Philip, 122
Peebles, P. James E. "Jim"
   Big Bang theory and, 139, 153
   CMBR and, 132
   cosmological constant and, 190
   flatness problem and, 153
   isotropic universe and, 152–153
   multiverse and, 190, 258
   Robert Dicke and, 132, 151–153
Peiris, Hiranya, 21, 183–187, *184*
pendulums, 39–40
Penrose, Lionel, 231
Penrose, Oliver, 231
Penrose, Roger
   Anthropic Principle and, 159–160
   Big Bang theory and, 160
   birth of, 231
   closed trapped surfaces and, 138
   Conformal Cyclic Cosmology and, 24, 217, 220, 231–237
   cyclic cosmologies and, 23–24, 217–220, 230–236
   early years of, 230–231
   general relativity and, 138
   gravitational theory and, 159
   as Nobel laureate, 138, 231
   photo of, *231*
   Stephen Hawking and, 138, 184
   Strong Anthropic Principle and, 159–160
   wave function and, 122
   Weyl Curvature Hypothesis and, 160
Penzias, Arno, 132, 140–141
Perlmutter, Saul, 209
perturbations, 154, 180, 229–230, 233
Petersen, Aage, 97–98, 103
phase space, 42–43, 54
Philolaus, 29, 33
photons, 21, 139–142, 169–171, 196, 207
physics, laws of, 23, 124, 238, 249, 254–255
Physics Letters (magazine), 179
Physics Reports (magazine), 149
Physics Today (magazine), 10, 109–110
Pi, So-Young, 178
Planck, Max, 70–71, 162
Planck length, 70–71, 145–146
Planck satellites, 141, 178, 185–186, 234–236, 270
Planck scale, 70–71, 145–146, 193, 200–207, 216, 220, 222–224
planetary orbits, 33–34, 41
planetary system, 19, 33–34, 157, 177, 211
planets, 22, 25, 29, 48, 69, 155–157
Plato, 28, 75
Poincaré, Henri, 44–47, *45*
Poincaré recurrence time, 42–47
"poker-playing," 73–76
polarization, 21, 122, 185–187, 274–275
Polchinski, Joseph, 207, 210
Popper, Karl, 254, 270
position, energy of, 39–40
positrons, 99, 150, 235–236
potential energy, 39–40, 175–181. *See also* energy
Power, Edwin, 105
Presley, Elvis, 259
Princeton University, 8, 76–78, 83, 90–98, 103–105, 110, 124, 129, 131, 141, 146, 152, 155, 171, 195, 198, 201–202, 212, 225, 246, 249
Principle of Least Action, 95
"Prisoner's Dilemma," 93
probability distribution, 77, 119
"Probability in Wave Mechanics," 102
probability waves, 85–86
Proceedings of the Prussian Academy of Sciences (journal), 66–67
Prussian Academy of Sciences, 64
psychic phenomena, 53–55
Ptolemy, 33–34

INDEX

pulsars, 137
Pythagoreans, 29, 33, 200

quantum chromodynamics (QCD), 197, 200–201
quantum electrodynamics (QED), 94, 97
quantum entanglement, 114–117, 191–217
quantum fluctuations, 160, 177–187, 208, 228, 234
quantum foam, 18, 145–146, 149, 188, 192, 220, 234–235
quantum measurement, 7–9, 76–81, 93–102, 111–123, 191–192, 260, 275
quantum mechanics, 5–10, 68–77, 80–91, 97–118, 154–158, 172–177, 239, 275
"Quantum Mechanics and Reality: Could the Solution to the Dilemma of Indeterminism Be a Universe in Which All Possible Outcomes of an Experiment Actually Occur?," 109
quantum processes, 6–8, 84–96, 115–119, 263–264
Quantum Simulator for Fundamental Physics (qSimFP), 187
quantum spin, 69–70, 120
quantum state, 6–10, 15, 84–88, 110–122, 191, 196, 275
quantum theory
 decoherence theory and, 116
 general relativity and, 99, 241–242
 of gravitation, 18, 108–109, 145, 170, 197–198, 234, 275
 wave function collapse and, 112
quantum tunneling, 173–176
quarks
 antiquarks and, 193–197
 electrons and, 120, 207
 gluons and, 197, 200
 Higgs mechanism and, 169–171
 leptons and, 164, 188, 193
Queen Mary College, 161, 203

radiation
 background radiation, 21, 129–133, 139–143, 148–151, 163–168, 174–178, 183–188, 220, 224, 227, 232–236, 260, 270, 274–275
 electromagnetic radiation, 50–51
 gravitational radiation, 170
 Hawking radiation, 135, 232, 234, 272

 infrared radiation, 1
 thermal radiation, 233
 radio waves, 132, 151, 273
Ramond, Pierre, 196
RAND Corporation, 93
Randall, Lisa, 222
Randall-Sundrum model, 223
Raychaudhuri, Amal Kumar, 138
rebirth concept, 219–239
recurrence time, 36, 42–47
reduction mechanisms, 121–122, 275
Rees, Martin, 24, 184, 214, 260
reheating process, 173–176
Reissner, Hans, 136–137
Reissner-Nordström black holes, 136–137
"Relative State Formulation of Quantum Mechanics," 106
religion, 33, 53, 92–94, 154, 219
"Remark About the Relationship Between Relativity Theory and Idealistic Philosophy," 240
Renaissance, 33–34
renormalization, 197–204, 276
replica Earths, 32–37, 46
replica people, 36, 107, 116–117, 121
replica worlds, 27, 32–37, 45–46, 100–101
Reviews of Modern Physics (journal), 105
Revolutionary War, 266
Rick and Morty (TV series), 2, 261–262, 266–267
Riemann, Bernhard, 43
Riess, Adam, 209
Rimini, Alberto, 122
Robertson, Howard P., 131–132, 147
Robinson, Kim Stanley, 252
Roosevelt, Franklin, 13
Rosen, Nathan, 146
"Rotating Cylinders and the Possibility of Global Causality Violation," 244
rotational symmetry, 39, 172–173. *See also* symmetry
Rovelli, Carlo, 234
rubber bands, 181, 193–197
Russell, Ken, 127
Rutgers University, 195, 207

Sacred Timeline, 262
Sagan, Carl, 184, 245–247
Salam, Abdus, 171, 273

INDEX

SAP (Strong Anthropic Principle), 19, 124, 157–165, 193, 212
Saturn, 25, 29, 33, 48
Saunders, Simon, 120
scalars
 explanation of, 49, 133
 Higgs scalar field, 169–171, 178
 scalar field, 49, 133, 151, 169–171, 173–181, 203, 215, 274
scale-invariant imprint, 178, 187–188, 224, 229
scattering profile, 94–95, 194
Scherk, Joël, 198–203
Schmidt, Brian, 209
Schrödinger, Erwin
 cat paradox and, 7–9, 85, 107, 115, 118
 equation by, 77, 85, 87, 98
 as Nobel laureate, 85–87
 photo of, 85
Schrödinger's cat, 7–9, 85, 107, 115, 118
Schrödinger's equation, 77, 85, 87, 98
Schwarz, John, 198–199, 199, 201–204
Schwarzschild, Karl, 104, 134, 136–137, 245
Sciama, Dennis, 138–139, 184, 230–231, 271
Scientific American (magazine), 22, 24, 228, 271
Scott, Douglas, 234
séances, 53–54, 63
Second Law of Thermodynamics, 44–46, 87, 115, 160
Seybert Commission, 55
Shakespeare, William, 62, 123, 267
Shechtman, Daniel, 225
Simons Observatory, 187
singularity, 128, 134–139, 148–152, 179, 231–234
six-dimensional space, 192, 195–198, 204–205
Slade, Henry, 54–55
"slate writing," 54
Sliders (TV series), 266
Sliding Doors (film), 265–266
"sliding doors moments," 264–266
Smolin, Lee, 135–136, 234
Society for Psychical Research, 54
Society of German Natural Scientists and Physicians, 62
solar eclipse, 63, 104, 185
solar system, 19, 25–27, 33–34, 69, 155–157, 177, 211, 246

Sommerfeld, Arnold, 69–70
"Sound of Thunder, A," 251
space, absolute, 17, 48, 56–61
space, abstract, 6–18, 42–43, 54, 83–84, 96, 146
space, phase, 42–43, 54
"Space Ripples Reveal Big Bang's Smoking Gun," 186
space-time, 5, 11–19, 31, 48, 56–64, 135, 145–146, 238, 244, 254
space-time foam, 145–146. *See also* quantum foam
spatial dimensions, 54–55, 147, 207. *See also* dimensions
special relativity
 Albert Einstein and, 11–12, 61–65, 195, 237–238, 243
 explanation of, 11–12, 237–238
 Herman Minkowski and, 12, 61–65
 time dilation and, 61, 243–245, 248–250
 time travel and, 244–250
 twin paradox and, 248–250
 see also general relativity
"Spectrum of Relict Gravitational Radiation and the Early State of the Universe," 170
speed of light, 17, 51–61, 132, 169, 191, 243–245, 270, 273
spheres, 44–45, 54, 128, 137
Spider-Man: No Way Home (film), 10, 269
spin vibrations, 194–196
Spinoza, Baruch, 74
spiritualism, 55, 60–63
spontaneous compactification, 199–200
spontaneous emergence of order, 160
spontaneous localization, 117, 121–123, 275
spontaneous symmetry, 172–173, 199–200. *See also* symmetry
Standard Model of particle physics, 120, 164, 171–172, 192–193, 201–208, 212, 222–223, 226
Stanford Linear Accelerator Center, 171
Stanford University, 171, 186, 208, 210, 231, 260
Star Trek (TV series), 261, 269
Starobinsky, Alexei, 170–171, 179, 224
Starobinsky inflation, 170–171
stars, 19, 25–57, 141, 232–233, 270
Steinhardt, Paul
 brane worlds and, 217, 220–224, 226–227, 230, 237

INDEX

bubble universes and, 183, 187
cyclic cosmologies and, 23–24, 219–220, 224–233, 237
Cyclic Universe and, 23–24, 220, 225–228
early years of, 224–225
Ekpyrotic Universe and, 23–24, 220, 224–229
eternal inflation and, 21–24, 179–183, 216–217, 224–230, 270–272
inflationary universe and, 20, 22–24, 224–230
Many Worlds Interpretation and, 166, 258
M-theory and, 217, 226
new inflation and, 175
photo of, 225
quantum fluctuations and, 179–180
string theory and, 216–217, 270–271
Stewart, Balfour, 53–54
Stewart, Patrick, 267
string landscape multiverse
Anthropic Principle and, 274
concept of, 14–21, 60, 188–189, 223
explanation of, 14–21
higher-dimensional spaces and, 191–217
string tension, 200
string theory
concept of, 3, 14–19, 270–271
configurations, 15–19
explanation of, 3, 14–19, 193–197
hadronic string theory, 194–197
higher-dimensional spaces and, 191–217, 220–228
inflationary universe and, 192–193, 210–217, 224–234
issues in, 165
John Wheeler and, 216
M-theory and, 14, 60, 72, 193, 202–210, 215–217, 221, 226–228, 234–237, 273–276
Paul Steinhardt and, 216–217, 270–271
quantum entanglement and, 191–217
string landscape multiverse and, 14–21, 60, 188–189, 191–217, 223, 270–276
superstring theory, 3, 14, 72–73, 164–165, 196–208, 215–217, 275
strings
closed strings, 204–207, 222
explanation of, 14, 193–197
open strings, 204–207, 222
rubber bands and, 181, 193–197

see also string theory
Strong Anthropic Principle (SAP), 19, 124, 157–165, 193, 212. *See also* Anthropic Principle
Stukeley, William, 47–48
subatomic world, 14, 88–89, 96, 115, 188, 193–197
"sum over histories," 6, 18, 94–101, 146
Sundrum, Raman, 222–223
super-cooling process, 168, 172–175
supergravity, 14, 72, 164, 201–208
supernatural beliefs, 53–54, 63
supernova, 209, 232
superpowers, 196–204
superspace, 146–149
superstring theory
dualities and, 205–206
Edward Witten and, 201–206, 211
explanation of, 3, 14
higher-dimensional spaces and, 164–165, 198–208
M-theory and, 72–73, 206, 215–217
revolutions and, 203–207
supersymmetry and, 14, 196–204, 216, 275
see also string theory
supersymmetry, 14, 196–204, 216, 275
survival of fittest, 136, 192, 207–215
Susskind, Leonard, 194–195, 209–213, 223
Swiss Federal Institute of Technology (ETH), 61–62
symmetry
conservation laws and, 38–46
endless loops and, 38–44
of nature, 38–41
rotational symmetry, 39, 172–173
spherical symmetry, 137
spontaneous symmetry, 172–173, 199–200
supersymmetry, 14, 196–204, 216, 275
time symmetry, 39
translational symmetry, 39
Syracuse University, 203

tachyonic cuts, 195
tachyons, 195, 244–245
Tait, Peter, 53, 55–56
Taub, Abraham, 147
Tegmark, Max, 11, 214
Templeton Prize, 271
ten-dimensional theory, 14, 198–200, 204–206

305

INDEX

tensors
    explanation of, 49–50, 65–66, 160
    five-dimensional tensors, 65–66
    four-dimensional tensors, 65–66
    of general relativity, 65–66
    tensor field, 49–50
    Weyl tensor, 160, 232
tesseract, *43*, 44
Theodicy (book), 82
"theory of everything," 205, 215, 258, 276
"Theory of the Universal Wave Function, The," 103
Theosophists, 55
thermodynamics, 44–46, 87, 115, 160
Thorne, Kip
    as author, 246
    early years of, 245–246
    geons and, 146, 246
    gravitational theory and, 237
    as Nobel laureate, 246
    Stephen Hawking and, 253–254
    time travel and, 12, 237, 246–254
    wormholes and, 12, 246–251
three dimensions, 11, 14–17, *43*, 43–44, 54–55, 61, 72–73, 128, 221
three loop contributions, 203
three-dimensional objects, *43*, 44, 54, 72, 145–146
three-dimensional space, 11, 14, 72–73, 221
time
    absolute time, 17, 48, 56–61
    cycles of, 219, 239–241, 255–256
    dilation of, 61, 243–245, 248–250
    flow of, 237–241, 254–256
    nature of, 237–241, 254–256
    recurrence time, 36, 42–47
    space-time, 5, 11–19, 31, 48, 56–64, 135, 145–146, 238, 244, 254
    space-time foam, 145–146
    symmetry and, 39
time dilation, 61, 243–245, 248–250
time symmetry, 39
time travel
    backward time travel, 13, 235–238, 242–244, 248–255
    branching timelines and, 238, 243, 250–256
    closed loops and, 244, 250–254
    concept of, 12–13, 215, 237–255
    flow of time and, 237–241, 254–256
    forward time travel, 247, 250
    Kip Thorne and, 12, 237, 246–254
    special relativity and, 244–250
    speed of light and, 243–245
    Stephen Hawking and, 240–241, 253–254
    time tunnels and, 248–249
    wormholes and, 12–13, 238, 245–248, 266
time tunnels, 248–249
Time Variance Authority (TVA), 262
time warps, 12, 63
Tipler, Frank, 158, 178, 244–245
Tipler cylinder, 244–245
Tod, Paul, 231–232
Tolman, Richard, 133, 219
torus, 72–73, 200. *See also* doughnut shapes
Transcendental Physics (book), 55
translational symmetry, 39. *See also* symmetry
Trimble, Virginia, 24, 218, 272
Trinity College, 213, 240
Trotter, Hale, *91*, 94
Tucker, Albert, 90, 93
Tufts University, 180
tunnels, quantum, 173–176
tunnels, time, 248–249
Turing, Alan, 254
Turok, Neil, 23, 220, 226–227, 232, 235–237
twin paradox, 248–250
twins, 248–250, 256, 261–264
two dimensions, *43*, 44, 75, 206
Tyndall, John, 41, 52–53, 272

Uhlenbeck, George, 69, 71–72
uncertainty principle, 5, 88–89, 95–96
"Unified Dynamics for Microscopic and Macroscopic Systems," 122
unified theory, 64, 70–71, 104, 134, 273–276
universal expansion, 131, 235. *See also* expanding universe
universal wave function, 9, 78, 100–103, 109–110, 122, 128–129, 192, 275
Universe or Multiverse? (book), 213
University College London, 183–184, 231
University of Berlin, 64
University of California, Berkeley, 212
University of California, Santa Barbara, 111, 205, 210
University of Cambridge, 94, 128–129, 138, 153, 176, 179, 226, 231, 240, 254, 271
University of Cape Town, 271

INDEX

University of Chicago, 194
University of Florida, 196
University of Göttingen, 59, 65
University of Königsberg, 64
University of London, 161, 203
University of Maryland, 146, 149
University of Michigan, 69
University of North Carolina, 104–105, 108, 146, 202
University of Notre Dame, 94
University of Oxford, 94, 111, 119–120, 231
University of Pennsylvania, 55, 175, 205, 225, 229
University of Pittsburgh, 119, 137
University of Southern California, 206
University of Texas, Austin, 110–111, 144, 205, 212
Unseen Universe, The (book), 53

vacuum cosmology, 148
vacuum speed of light, 51. *See also* speed of light
van Nieuwenhuizen, Peter, 201
van Tent, Bartjan, 235
vector field, 49–50
Veneziano, Gabriele, 194
Venus, 29
"verse-jumping," 263
Very Early Universe workshop, 176–177
vibrations, 194–196, 200
Vilenkin, Alexander, 20, 180–181, 187
Visser, Matt, 247
Voltaire, 16
von Neumann, John, 6–7, 9, 77, 79, *83*, 83–90, 92–93, 97, 115

Waldron, Michael, 262
Walker, Arthur, 131–132
Wallace, Alfred Russel, 53
Wallace, David, 119–120
WAP (Weak Anthropic Principle), 156, 158, 163
warps, 12, 63
wave equation, 50, 77, 85–86, 273
wave function, 9, 76–78, 85–90, 98–117, 121–122, 128–129, 192, 275
wave function collapse, 9, 77–78, 86–90, 98–112, 117
WCH (Weyl Curvature Hypothesis), 160, 231–232

Weak Anthropic Principle (WAP), 156, 158, 163
Webb, James, 177
Webb space telescope, 1–2, 25, 141, 177
Weber, Joseph, 105
Weber, Tullio, 122
Weber, Wilhelm, 53
Weinberg, Steven
  Anthropic Principle and, 193, 208–214, 223
  birth of, 211
  cosmological constant and, 208
  early years of, 211–212
  electromagnetism and, 171
  electroweak unification and, 273
  higher-dimensional spaces and, 191, 193, 204, 208–214
  multiverse concept and, 260
  as Nobel laureate, 188, 193, 204, 212
  photo of, *212*
  Standard Model of particle physics and, 212
  Strong Anthropic Principle and, 193, 212
Weiss, Rainer, 246
Wellington, Duke of, 36
Wess, Julius, 198
Weyl Curvature Hypothesis (WCH), 160, 231–232
Weyl tensor, 160, 232
What If...? (TV series), 262
Wheeler, John
  Albert Einstein and, 77–78
  Anthropic Principle and, 129, 134, 148–151, 158
  as author, 246
  Brandon Carter and, 155–158
  Charles Misner and, 78, 94, 101–104
  general relativity and, 77–78, 94–99, 104–105
  gravitational theory and, 18–19, 104
  Hugh Everett and, 8–9, 97–104, 106–110, 144
  Niels Bohr and, 97–104, 109–110
  Participatory Universe idea, 110–112
  photo of, *95*
  quantum electrodynamics and, 94, 97
  quantum foam and, 18, 145–146, 149, 188, 234–235
  radical conservatism philosophy, 99
  string theory and, 216

INDEX

Wheeler, John (*continued*)
   wave function and, 98–99
   wormholes and, 12
white holes, 245
"Why Is the Universe Isotropic?," 153
Wigner, Eugene, 97
Wilkinson Microwave Anisotropy Probe (WMAP), 21, 141, 178, 183–185, 220, 233
William of Ockham, 112
Wilson, Robert, 132, 140–141
Witten, Edward
   early years of, 201–202
   general relativity and, 201–202
   gravitational theory and, 201–202
   higher-dimensional spaces and, 191, 201–206, 211
   M-theory and, 202, 206
   photo of, *202*
   supergravity and, 201–206
   superstring theory and, 201–206, 211
Witten, Louis, 201–202
WMAP (Wilkinson Microwave Anisotropy Probe), 21, 141, 178, 183–185, 220, 233
Woit, Peter, 212–213
World War I, 65, 134
World War II, 251–252
wormholes
   explanation of, 12–13, 146, 238
   John Wheeler and, 12
   Kip Thorne and, 12, 246–251
   Michael Morris and, 246–251
   stable wormholes, 246–247
   time travel and, 12–13, 238, 245–248, 266
   traversable wormholes, 12–13, 238, 245–248, 266

X2 (film), 267
X-Men United (film), 267

Yau, Shing-Tung, 73, 205
Years of Rice and Salt, The (book), 252
Yorke, James, 149
Yurtsever, Ulvi, 248–251

Zeh, H. Dieter, 114–115, 121
Zeilinger, Anton, 114
zero curvature, 129, 152
Zhurnal Éksperimental'noĭ i Teoreticheskoĭ Fiziki (journal), 170
Zierler, David, 166, 258
Zöllner, Johann, 54–55
Zumino, Bruno, 198, 201
Zur Theorie der Gesellschaftsspiele (On the Theory of Parlor Games) (paper), 93
Zurek, Wojciech, 116

Credit: Saint Joseph's University

Paul Halpern is a professor of physics at Saint Joseph's University and the author of eighteen popular science books, including *Flashes of Creation*, *The Quantum Labyrinth*, *Einstein's Dice and Schrödinger's Cat*, and *Synchronicity*. He is the recipient of a Guggenheim fellowship and is a fellow of the American Physical Society. He lives near Philadelphia, Pennsylvania.